Jesús M. Ruiz

The Basic Theory of Power Series

Advanced Lectures
in Mathematics

Editorial board:
Prof. Dr. Martin Aigner, Freie Universität Berlin, Germany
Prof. Dr. Gerd Fischer, Heinrich-Heine-Universität Düsseldorf, Germany
Prof. Dr. Michael Grüter, Universität des Saarlandes, Saarbrücken, Germany
Prof. Dr. Manfred Knebusch, Universität Regensburg, Germany
Prof. Dr. Gisbert Wüstholz, ETH Zürich, Switzerland

Jesús M. Ruiz
The Basic Theory of Power Series

Heinrich von Weizsäcker and Gerhard Winkler
Stochastic Integrals

Francesco Guaraldo, Patrizia Macri, and Alessandro Tancredi
Topics on Real Analytic Spaces

Manfred Denker
Asymptotic Distribution Theory in Nonparametric Statistics

Jochen Werner
Optimization. Theory and Applications

Jesús M. Ruiz

The Basic Theory of Power Series

Prof. Dr. Jesús M. Ruiz
Departamento de Geometría y Topología
Facultad de Matemáticas
Universidad Complutense de Madrid
28040 Madrid, Spain

AMS Subject Classification: 13-01, 13F20, 13F25, 13J05

All rights reserved
© Friedr. Vieweg & Sohn Verlagsgesellschaft mbH, Braunschweig/Wiesbaden, 1993

Vieweg is a subsidiary company of the Bertelsmann Publishing Group International.

No part of this publication may be reproduced, stored in a retrieval system or transmitted, mechanical, photocopying or otherwise, without prior permission of the copyright holder.

Cover design: Klaus Birk, Gießen
Printing and binding: W. Langelüddecke, Braunschweig
Printed on acid-free paper
Printed in Germany

ISSN 0932-7134
ISBN 3-528-06525-7

To MariPaz

Preface

The aim of these notes is to cover the basic algebraic tools and results behind the scenes in the foundations of Real and Complex Analytic Geometry. The author has learned the subject through the works of many mathematicians, to all of whom he is indebted. However, as the reader will immediately realize, he was specially influenced by the writings of S.S. Abhyankar and J.-C. Tougeron. In any case, the presentation of all topics is always as elementary as it can possibly be, even at the cost of making some arguments longer. The background formally assumed consists of:

1) Polynomials: roots, factorization, discriminant; real roots, Sturm's Theorem, formally real fields; finite field extensions, Primitive Element Theorem.

2) Ideals and modules: prime and maximal ideals; Nakayama's Lemma; localization.

3) Integral dependence: finite ring extensions and going-up.

4) Noetherian rings: primary decomposition, associated primes, Krull's Theorem.

5) Krull dimension: chains of prime ideals, systems of parameters; regular systems of parameters, regular rings.

These topics are covered in most texts on Algebra and/or Commutative Algebra. Among them we choose here as general reference the following two:

- M. Atiyah, I.G. Macdonald: Introduction to Commutative Algebra, 1969, Addison-Wesley: Massachusetts; quoted [A-McD].

- S. Lang: Algebra, 1965, Addison-Wesley: Massachusetts; quoted [L].

Even quotations to these two books are kept to a minimum, avoiding them whenever a reasonably self-contained explanation could be provided. In this way many deep results can be obtained for rings of power series almost from scratch, giving them a highly geometrical meaning. Two examples of this are that localizations preserve regularity or that integral closures are finite. In fact, that is one of the goals of this book: to review the commutative algebra of power series in geometrical terms. The guidelines for this review are the Local Parametrization Theorem, the Nullstellensatz, and Zariski's Main Theorem. Furthermore, all the work is carried out both in the complex and the real case, showing the additional difficulties and the peculiarities of the latter.

All in all, the final hope is that this book will be of some help to those not acquainted with either the geometry behind local commutative algebra or the algebra behind local analytic geometry.

The notes are based on courses and seminars given by the author during many years, organized by the Departments of Algebra, Geometry, and Topology, at the Universidad Complutense de Madrid, with the support of the D.G.I.C.Y.T.. Several people have made these activities possible and enthralling, and have contributed to the actual writing of the book. First of all, Tomás Recio, who led the author into the beauty of Analytic Geometry. Also, Carlos Andradas, MariEmi Alonso, and José Manuel Gamboa, who provided enjoyable and fruitful discussions on the teaching and learning of power series, and Capi Corrales, who read word by word the first draft of the manuscript and suggested all kind of accurate corrections.

The final thanks are for MariPaz, to whom this book, and everything else, is dedicated.

Majadahonda, October 1992

Contents

I	**Power Series**	**1**
1	Series of Real and Complex Numbers	1
2	Power Series	3
3	Rückert's and Weierstrass's Theorems	11
II	**Analytic Rings and Formal Rings**	**16**
1	Mather's Preparation Theorem	16
2	Noether's Projection Lemma	21
3	Abhyankar's and Rückert's Parametrization	26
4	Nagata's Jacobian Criteria	31
5	Complexification	39
III	**Normalization**	**45**
1	Integral Closures	45
2	Normalization	49
3	Multiplicity in Dimension 1	53
4	Newton-Puiseux's Theorem	58
IV	**Nullstellensätze**	**64**
1	Zero Sets and Zero Ideals	64
2	Rückert's Complex Nullstellensatz	68
3	The Homomorphism Theorem	73
4	Risler's Real Nullstellensatz	78
5	Hilbert's 17th Problem	81
V	**Approximation Theory**	**87**
1	Tougeron's Implicit Functions Theorem	87
2	Equivalence of Power Series	90
3	M. Artin's Approximation Theorem	94
4	Formal Completion of Analytic Rings	99
5	Nash Rings	106
VI	**Local Algebraic Rings**	**110**
1	Local Algebraic Rings	110
2	Chevalley's Theorem	112
3	Zariski's Main Theorem	115
4	Normalization and Completion	120
5	Efroymson's Theorem	125
Bibliographical Note		**130**
Index		**133**

I   Power Series

Summary. In this chapter we study in detail convergent and formal power series. We start by recalling the notion of convergence and its main properties: absolute convergence and reordering of convergent series. We then introduce formal and convergent series and discuss the operations with them: sum, multiplication, substitution and derivation. In particular we get the Identity Principle for functions associated to convergent power series. Finally we state and prove three essential results of the theory: Rückert's Division Theorem, Weierstrass's Preparation Theorem and Hensel's Lemma.

1   Series of Real and Complex Numbers

We will denote by $\sum_\nu a_\nu$ or $\sum a_\nu$ a series of elements of the field $\mathbb{K} = \mathbb{R}$ or $\mathbb{C}$. Here the indices $\nu = (\nu_1, \ldots, \nu_n)$ are elements of $\mathbb{N}^n$, and we will use the standard notations $|\nu| = \nu_1 + \cdots + \nu_n$ and $\nu! = \nu_1! \cdots \nu_n!$.

Definition 1.1 *The series $\sum a_\nu$ converges to the element $c \in \mathbb{K}$ if for every real number $\varepsilon > 0$ there is a finite set $I_\varepsilon \subset \mathbb{N}^n$ such that $|\sum_{\nu \in I} a_\nu - c| < \varepsilon$ for every finite set of indices $I \supset I_\varepsilon$. In that case we say that c is the sum of the series and we write $c = \sum a_\nu$.*

Remarks 1.2 *a)* For series with indices in $\mathbb{N}$ the convergence in the sense of Definition 1.1 implies the classical one, but not conversely: $\sum (-1)^k / k$ does not converge according to 1.1, nevertheless the limit $\lim_{p \to \infty} \sum_{1 \leq k \leq p} (-1)^k / k$ does exist.
  b) If $\sum a_\nu$ converges to $c \in \mathbb{K}$, then $c = \lim_{p \to \infty} \sum_{|\nu| \leq p} a_\nu$.
  c) Let $\sum a_\nu$, $\sum b_\nu$ converge to $c, d \in \mathbb{K}$ respectively, and consider $\lambda, \mu \in \mathbb{K}$. Then $\sum (\lambda a_\nu + \mu b_\nu)$ converges to $\lambda c + \mu d$.
  d) Let $\sum a_\nu$, $\sum b_\nu$ be two convergent series of real numbers such that $\sum_{\nu \in I} a_\nu \leq \sum_{\nu \in I} b_\nu$ for every finite set of indices I. Then $\sum a_\nu \leq \sum b_\nu$.

Proposition 1.3 *A series $\sum a_\nu$ of non-negative real numbers converges if and only if there is $M > 0$ such that $\sum_{\nu \in I} a_\nu < M$ for every finite set of indices $I \subset \mathbb{N}^n$. In this case the sum of the series is the supremum of all those finite sums.*

Proof. If there is such an M, then the supremum $c = \sup \{ \sum_{\nu \in I} a_\nu \mid I \text{ finite} \}$ exists. Let us see that the series converges to c. Indeed, for every $\varepsilon > 0$ there is a finite set of indices I_ε such that $c - \sum_{\nu \in I} a_\nu < \varepsilon$, and for every finite set $I \supset I_\varepsilon$ we have

$$c \geq \sum_{\nu \in I} a_\nu \geq \sum_{\nu \in I_\varepsilon} a_\nu,$$

so that

$$0 \leq c - \sum_{\nu \in I} a_\nu \leq c - \sum_{\nu \in I_\varepsilon} a_\nu < \varepsilon.$$

Conversely, supose that $\sum a_\nu$ converges to c. Then $c = \lim_{p \to \infty} \sum_{|\nu| \leq p} a_\nu$ (Remark 1.2 b)), and the a_ν's being non-negative, we can take $M = c$. □

Proposition 1.4 *A series $\sum a_\nu$ of real numbers converges if and only if the series $\sum |a_\nu|$ converges. In this case $|\sum a_\nu| \leq \sum |a_\nu|$.*

Proof. For every ν we set $p_\nu = \max\{a_\nu, 0\}$, $q_\nu = \max\{-a_\nu, 0\}$.

If the series $\sum |a_\nu|$ converges, so do $\sum p_\nu$ and $\sum q_\nu$ by Proposition 1.3. Hence also the series $\sum (p_\nu - q_\nu)$ converges (Remark 1.2 c)), and since $p_\nu - q_\nu = a_\nu$, the series $\sum a_\nu$ converges, too.

Now suppose that $\sum a_\nu$ converges to c. Take $\varepsilon = 1$ and consider the set of indices I_ε provided by the definition of convergence. For every $L \subset \mathbb{N}^n$ write

$$L^+ = \{\nu \in L \mid a_\nu > 0\}, \qquad L^- = \{\nu \in L \mid a_\nu \leq 0\}.$$

With these notations, for every finite set of indices I we put $J = I^+ \cup I_\varepsilon$, and since the p_ν's are all non-negative, $J \supset I_\varepsilon$ and $J^- = I_\varepsilon^-$ we have

$$\sum_{\nu \in I} p_\nu = \sum_{\nu \in I^+} p_\nu \leq \sum_{\nu \in J} p_\nu = \sum_{\nu \in J^+} a_\nu = \sum_{\nu \in J} a_\nu - \sum_{\nu \in J^-} a_\nu \leq$$
$$\leq |\sum_{\nu \in J} a_\nu - c| + |\sum_{\nu \in J^-} a_\nu - c| < \varepsilon + |\sum_{\nu \in I_\varepsilon^-} a_\nu - c|.$$

Hence by Proposition 1.3 the series $\sum p_\nu$ converges. Analogously, $\sum q_\nu$ converges, and by Remark 1.2 c) so does the series $\sum (p_\nu + q_\nu)$. As $|a_\nu| = p_\nu + q_\nu$ we are done. The last inequality of the statement follows from Remark 1.2 b). □

We will denote by $\Re(z)$ (resp. $\Im(z)$) the real (resp. imaginary) part of a complex number $z \in \mathbb{C}$. We clearly have

$$|z| \leq |\Re(z)| + |\Im(z)|, \ |\Re(z)| \leq |z|, \ |\Im(z)| \leq |z|.$$

From these inequalities and the preceding propositions one easily gets:

Proposition 1.5 *Let $\sum a_\nu$ be a series of complex numbers. Then:*

a) *$\sum a_\nu$ converges if and only if the two series $\sum \Re(a_\nu)$ and $\sum \Im(a_\nu)$ converge.*

b) *$\sum |a_\nu|$ converges if and only if the two series $\sum |\Re(a_\nu)|$ and $\sum |\Im(a_\nu)|$ converge.*

c) $\sum a_\nu$ converges if and only if $\sum |a_\nu|$ converges. In this case, $|\sum a_\nu| \leq \sum |a_\nu|$.

Proposition 1.6 *Let $\sum a_\nu$ be a convergent series and consider a permutation τ of $\{1,\ldots,n\}$. Then the sum of the iterated series $\sum_{\nu_{\tau(1)}} \cdots \sum_{\nu_{\tau(n)}} a_\nu$ converges and coincides with the sum of $\sum a_\nu$.*

Proof. We will suppose for simplicity that τ is the identity, and argue by induction on n. For $n=1$ the assertion is trivial, so that we assume $n>1$, and the result is true for series with indices in $\mathbb{N}^{n-1}$.

For every $\nu_1 \in \mathbb{N}$ the series $\sum_{\nu'} a_{\nu_1 \nu'}$ converges. Indeed, let us see that $\sum_{\nu'} |a_{\nu_1 \nu'}|$ converges. Consider a finite set of indices $I' \subset \mathbb{N}^{n-1}$ and note that

$$\sum_{\nu' \in I'} |a_{\nu_1 \nu'}| = \sum_{\nu \in \{\nu_1\} \times I'} |a_{\nu_1 \nu'}|.$$

By Proposition 1.5 the series of non-negative real numbers $\sum |a_\nu|$ converges, and consequently it verifies the criterion of Proposition 1.3. By the above equality, the series $\sum_{\nu'} |a_{\nu_1 \nu'}|$ also verifies that criterion, and hence converges. It follows that $\sum_{\nu'} a_{\nu_1 \nu'}$ converges, too, and we write $b_{\nu_1} = \sum_{\nu'} a_{\nu_1 \nu'}$. Now, by the induction hypothesis, b_{ν_1} is also the sum of the iterated series $\sum_{\nu_2} \cdots \sum_{\nu_n} a_\nu$.

To conclude the proof, we have to show that the series $\sum b_{\nu_1}$ converges to $c = \sum a_\nu$. Thus let $\varepsilon > 0$. We pick a finite set of indices $I_{\varepsilon/2} \subset \mathbb{N}^n$ such that $|\sum_{\nu \in I} a_\nu - c| < \varepsilon/2$ for every finite set $I \supset I_{\varepsilon/2}$ (convergence of $\sum a_\nu$), and we let $I_\varepsilon^{(1)}$ be the set of all the first components of the indices in $I_{\varepsilon/2}$. Then for every finite set $I^{(1)} \supset I_\varepsilon^{(1)}$ with m elements we choose a finite set $I' \subset \mathbb{N}^{n-1}$ such that $I_{\varepsilon/2} \subset I^{(1)} \times I'$ and

$$|\sum_{\nu' \in I'} a_{\nu_1 \nu'} - b_{\nu_1}| < \frac{1}{2m}\varepsilon$$

for every $\nu_1 \in I^{(1)}$. We then have:

$$|\sum_{\nu_1 \in I^{(1)}} b_{\nu_1} - c| = |\sum_{\nu_1 \in I^{(1)}} b_{\nu_1} - \sum_{\nu_1 \in I^{(1)}} \sum_{\nu' \in I'} a_{\nu_1 \nu'} + \sum_{\nu_1 \in I^{(1)}} \sum_{\nu' \in I'} a_{\nu_1 \nu'} - c| \leq$$

$$\leq \sum_{\nu_1 \in I^{(1)}} |b_{\nu_1} - \sum_{\nu' \in I'} a_{\nu_1 \nu'}| + |\sum_{\nu \in I^{(1)} \times I'} a_\nu - c| < m(\varepsilon/2m) + \varepsilon/2 = \varepsilon.$$

This shows that the series $\sum b_{\nu_1}$ converges to c, as wanted. □

2 Power Series

The affine space $\mathbb{K}^n$ (where $\mathbb{K} = \mathbb{R}$ or $\mathbb{C}$) will always be endowed with the usual euclidean topology. As is well known, each point $x_0 = (x_{01},\ldots,x_{0n}) \in \mathbb{K}^n$ has a neighborhood basis that consists of the polycylinders Δ of polyradius $\rho = (\rho_1,\ldots,\rho_n)$, $\rho_i > 0$, centered at x_0:

$$\Delta = \{x \in \mathbb{K}^n \mid |x_i - x_{0i}| < \rho_i \text{ for } 1 \le i \le n\}.$$

Definitions and Notations 2.1 A *formal power series* in the indeterminates $\mathbf{x}_1, \ldots, \mathbf{x}_n$ is an expression $f = \sum_{\nu \in \mathbb{N}^n} a_\nu \mathbf{x}_1^{\nu_1} \cdots \mathbf{x}_n^{\nu_n}$, in short $\sum_\nu a_\nu \mathbf{x}^\nu$ or $\sum a_\nu \mathbf{x}^\nu$, where $a_\nu \in \mathbb{K}$ for every ν; the a_ν's are the *coefficients* of $\sum a_\nu \mathbf{x}^\nu$, and the first of them $a_{(0,\ldots,0)}$ is denoted by $f(0)$. The set of all these formal power series will be denoted by $\mathcal{F}_n$, $\mathbb{K}[[\mathbf{x}_1, \ldots, \mathbf{x}_n]]$ or $\mathbb{K}[[\mathbf{x}]]$.

The *order* of a formal power series $\sum a_\nu \mathbf{x}^\nu$, denoted by $\omega(f)$, is the smallest integer $p \ge 0$ such that $a_\nu \ne 0$ for some ν with $|\nu| = p$, provided that some a_ν is $\ne 0$. Otherwise, that is, if $f = 0$, we write $\omega(f) = +\infty$.

Let $f = \sum a_\nu \mathbf{x}^\nu$ be a formal power series. If $x \in \mathbb{K}^n$ and the series $\sum a_\nu x^\nu$ of elements of $\mathbb{K}$ converges to $c \in \mathbb{K}$, we say that f *converges at x to c*, and write $f(x) = c$. Now let $D \subset \mathbb{K}^n$. We say that f *converges uniformly on D* if:

a) f converges at every point of D, and

b) for every $\varepsilon > 0$ there is a finite set $I_\varepsilon \subset \mathbb{N}^n$ such that $|\sum_{\nu \in I_\varepsilon} a_\nu x^\nu - f(x)| < \varepsilon$ for every finite set of indices $I \supset I_\varepsilon$ and every point $x \in D$.

Let f be a formal power series. The *domain* of f, denoted by $D(f)$, is the interior of the set of points at which f converges. The series f is called *convergent* if $D(f) \ne \emptyset$. The set of all convergent power series will be denoted by $\mathcal{O}_n$, $\mathbb{K}\{\mathbf{x}_1, \ldots, \mathbf{x}_n\}$ or $\mathbb{K}\{\mathbf{x}\}$.

For the time being we will use the following notation: Given a point $x^* = (x_1^*, \ldots, x_n^*) \in \mathbb{K}^n$ whose coordinates are none zero, $\Delta(x^*)$ will stand for the polycylinder of polyradius $\rho = (|x_1^*|, \ldots, |x_n^*|)$ centered at the origin.

Proposition 2.2 *Let $f = \sum a_\nu \mathbf{x}^\nu$ be a convergent power series and $D^*(f)$ the set of points at which f converges and whose coordinates are none zero. The set $D(f)$ is the union of the $\Delta(x^*)$, $x^* \in D^*(f)$. In particular $D(f)$ is an open connected neighborhood of the origin. Furthermore, f converges uniformly on every compact subset of $D(f)$.*

Proof. It is enough to show that for $0 < r < 1$ and $x^* \in D^*(f)$, the series f converges uniformly on $\Delta(rx^*)$. To that end, note first that $\sum |a_\nu x^{*\nu}|$ coverges because so does $\sum a_\nu x^{*\nu}$ (Proposition 1.5 c)). Then there is $M > 0$ such that $|a_\nu x^{*\nu}| < M$ for all ν (Proposition 1.3). Now, if $x \in \Delta(rx^*)$ and $I \subset \mathbb{N}^n$ is finite we have

$$\sum_{\nu \in I} |a_\nu x^\nu| \le \sum_{\nu \in I} |a_\nu x^{*\nu}| r^{|\nu|} < M \sum_{\nu \in I} r^{|\nu|}.$$

But since

$$\sum_{\nu \in I} r^{|\nu|} \le \left(\sum_{\nu_1 \in I_1} r^{\nu_1}\right) \cdots \left(\sum_{\nu_n \in I_n} r^{\nu_n}\right),$$

where $I_\ell \subset \mathbb{N}$ is the set of the ℓ-th components of all the indices of I, and the series $\sum r^{\nu_\ell}$, $1 \le \ell \le n$, converge for $0 < r < 1$, so do the two series $\sum_{\nu \in I} r^{|\nu|}$ and $\sum |a_\nu x^\nu|$ (Proposition 1.3 again). We hence conclude that the series $\sum a_\nu x^\nu$ converges.

2 Power Series

It remains to see that this convergence is uniform on $\Delta(rx^*)$. Thus consider $\varepsilon > 0$ and a finite set $I_\varepsilon \subset \mathbb{N}^n$ such that $\sum_{\nu \notin I} r^{|\nu|} < \varepsilon/M$ for every finite set of indices $I \supset I_\varepsilon$. From one of our preceding inequalities we get

$$\left|\sum_{\nu \in I} a_\nu x^\nu - f(x)\right| \leq \sum_{\nu \notin I} |a_\nu x^\nu| \leq M \sum_{\nu \notin I} r^{|\nu|} < \varepsilon,$$

which concludes the proof. □

We now consider new indeterminates $y_1, \ldots, y_n$ and for every $\mu \in \mathbb{N}^n$ the formula

$$(x_1 + y_1)^{\mu_1} \cdots (x_n + y_n)^{\mu_n} = \sum_\nu P_{\mu\nu}(x) y^\nu,$$

where $P_{\mu\nu}(x) = \frac{\mu!}{\nu!(\mu-\nu)!} x^{\mu-\nu}$ and $\mu_1 \geq \nu_1, \ldots, \mu_n \geq \nu_n$. Then:

Proposition and Definition 2.3 *Let $f = \sum a_\nu x^\nu$ be a convergent power series. Then the associated function*

$$^a f : D(f) \to \mathbb{K} : x \mapsto f(x)$$

is continuous, and for every $x_0 \in D(f)$ it holds:

a) For every ν the series $\sum_\mu a_\mu P_{\mu\nu}(x_0)$ converges, say $b_\nu = \sum_\mu a_\mu P_{\mu\nu}(x_0)$.

b) The power series $g = \sum b_\nu x^\nu$ is convergent, and $g(x - x_0) = f(x)$ for x close enough to x_0.

Proof. For every integer $p \geq 0$ let f_p stand for the polynomial $\sum_{|\nu| \leq p} a_\nu x^\nu$. Now fix $0 < r < 1$ and $x^* \in D^*(f)$. By Proposition 2.2 the sequence of polynomials $(f_p)_{p \geq 0}$ converges uniformly to $^a f|\Delta(rx^*)$ on $\Delta(rx^*)$. Hence $^a f|\Delta(rx^*)$ is continuous and, $D(f)$ being the union of all the $\Delta(rx^*)$'s, $^a f$ is continuous.

To prove a) and b) note that if $y \in \mathbb{K}^n$ is close enough to the origin, the point $z = (|x_{01}| + |y_1|, \ldots, |x_{0n}| + |y_n|)$ belongs to $D(f)$ (Proposition 2.2 again), and it follows then that the series $\sum_{(\mu,\nu)} a_\mu P_{\mu\nu}(x_0) y^\nu$ converges. Indeed, if I, J are finite sets we have

$$\sum_{(\mu,\nu) \in I \times J} |a_\mu P_{\mu\nu}(x_0) y^\nu| \leq \sum_{\mu \in I} |a_\mu| \sum P_{\mu\nu}(|x_{01}|, \ldots, |x_{0n}|) |y^\nu| = \sum_{\mu \in I} |a_\mu| z^\mu,$$

and our assertion follows from Proposition 1.3, and the fact that f converges at z. Thus, the iterated series

$$\sum_\nu \sum_\mu a_\mu P_{\mu\nu}(x_0) y^\nu, \quad \sum_\mu \sum_\nu a_\mu P_{\mu\nu}(x_0) y^\nu,$$

exist and their sums coincide (Proposition 1.6). Whence we have proved a) and that the series g of b) is convergent. Finally, if x is close to x_0, then $y = x - x_0$ is close to the origin, and by the remark above

$$g(x - x_0) = \sum_\nu \left(\sum_\mu a_\mu P_{\mu\nu}(x_0) \right) y^\nu =$$

$$= \sum_\mu \left(\sum_\nu a_\mu P_{\mu\nu}(x_0) y^\nu \right) = f(x_0 + y) = f(x).$$

$\square$

(2.4) Operations with power series. Let $f = \sum a_\nu \mathbf{x}^\nu$ and $f = \sum b_\nu \mathbf{x}^\nu$ be two formal power series. We define their *sum* by

$$f + g = \sum (a_\nu + b_\nu) \mathbf{x}^\nu$$

and their product by

$$fg = \sum \left(\sum_{\lambda + \mu = \nu} a_\lambda b_\mu \right) \mathbf{x}^\nu.$$

One easily checks that

$$(1 - \mathbf{x}_1) \cdots (1 - \mathbf{x}_n) \sum_{|\nu| \geq 0} \mathbf{x}^\nu = 1.$$

Clearly, it holds

$$\omega(f + g) \geq \min\{\omega(f), \omega(g)\}, \quad \omega(fg) = \omega(f)\omega(g).$$

In particular, if $f \neq 0$ and $g \neq 0$, then $fg \neq 0$. If f and g converge at the point $x \in \mathbb{K}^n$, then $f + g$ and fg converge at that point, and

$$(f + g)(x) = f(x) + g(x), \quad (fg)(x) = f(x)g(x).$$

In particular, if f and g are convergent power series,

$$D(f + g) \supset D(f) \cap D(g) \neq \emptyset, \quad D(fg) \supset D(f) \cap D(g) \neq \emptyset,$$

which implies that the series $f + g$ and fg are also convergent.

In this way, $\mathcal{F}_n$ (resp. $\mathcal{O}_n$) is a commutative ring with unit, which contains the field of coefficients $\mathbb{K}$, and hence it is a $\mathbb{K}$-algebra. Furthermore, it is an integral domain.

A family $\{f_\lambda \mid \lambda \in \Lambda\}$ of formal power series

$$f_\lambda = \sum a_{\lambda\nu} \mathbf{x}^\nu$$

is called *summable* if for every integer $p \geq 0$ the subfamily of the series of order $\leq p$ is finite. In that case, for every $\nu \in \mathbb{N}^n$ set of λ's with $a_{\lambda\nu} \neq 0$ is finite (if $a_{\lambda\nu} \neq 0$ then $\omega(f_\lambda) \leq |\nu|$), and the sum $\sum_{\lambda \in \Lambda} a_{\lambda\nu}$ is finite. Consequently, the formal power series $\sum \left(\sum_{\lambda \in \Lambda} a_{\lambda\nu} \right) \mathbf{x}^\nu$ is well defined: it is called the *sum* of the family $\{f_\lambda \mid \lambda \in L\}$ and denoted by $\sum f_\lambda$.

2 Power Series

If two families $\{f_\lambda \mid \lambda \in \Lambda\}$ and $\{g_\lambda \mid \lambda \in \Lambda\}$ are summable, and a, b are two formal power series, the family $\{af_\lambda + bg_\lambda \mid \lambda \in \Lambda\}$ is summable, and its sum is the series $a \sum f_\lambda + b \sum g_\lambda$.

(2.5) Substitution. Let $f = \sum a_\nu \mathbf{x}^\nu, g_1, \ldots, g_n$ be formal power series with orders $\omega(g_1) \geq 1, \ldots, \omega(g_n) \geq 1$. Then for every ν,

$$\omega(a_\nu g_1^{\nu_1} \cdots g_n^{\nu_n}) \geq \nu_1 \omega(g_1) + \cdots + \nu_n \omega(g_n) \geq |\nu|,$$

and so the family

$$\{a_\nu g_1^{\nu_1} \cdots g_n^{\nu_n} \mid \nu \in \mathbb{N}^n\}$$

is summable. The sum of this family is called the *substitution of* $g_1, \ldots, g_n$ *in* f and denoted by $f(g_1, \ldots, g_n)$.

It is a straightforward computation to check that for any other formal power series h it holds

a) $(h + f)(g_1, \ldots, g_n) = h(g_1, \ldots, g_n) + f(g_1, \ldots, g_n)$, and

b) $(hf)(g_1, \ldots, g_n) = h(g_1, \ldots, g_n) f(g_1, \ldots, g_n)$.

As an application, consider the identity

$$(1 - \mathbf{x}_1) \sum \mathbf{x}_1^\nu = 1.$$

Then for every $f \in \mathcal{F}_n$ with $f(0) = a \neq 0$, we get

$$1 = \left(1 - \left(1 - \frac{1}{a}f\right)\right) \sum \left(1 - \frac{1}{a}f\right)^{\nu_1},$$

and consequently there exists the formal power series

$$\frac{1}{f} = \frac{1}{a} \sum \left(1 - \frac{1}{a}f\right)^{\nu_1}.$$

It follows that $\mathcal{F}_n$ is a local ring whose maximal ideal $\widehat{\mathfrak{m}}_n$ consists of the formal power series with order ≥ 1. Clearly this ideal is generated by the indeterminates: $\widehat{\mathfrak{m}}_n = \{\mathbf{x}_1, \ldots, \mathbf{x}_n\} \mathcal{F}_n$.

Concerning convergence we have

Proposition 2.6 *If* $f, g_1, \ldots, g_n$ *are convergent, then* $f(g_1, \ldots, g_n)$ *is convergent. More explicitely, if* $x \in \mathbb{K}^n$ *is close to the origin, then* $g_1, \ldots, g_n, f(g_1, \ldots, g_n)$ *converge at* x, f *converges at* $(g_1(x), \ldots, g_n(x))$ *and*

$$f(g_1, \ldots, g_n)(x) = f(g_1(x), \ldots, g_n(x)).$$

Proof. Let $f = \sum a_\nu \mathbf{x}^\nu$, $g_i = \sum b_{i\nu}\mathbf{x}^\nu, 1 \leq i \leq n$, and write $g_i^* = \sum |b_{i\nu}|\mathbf{x}^\nu, 1 \leq i \leq n$. By Proposition 2.3 and since $g_1^*(0) = \cdots = g_n^*(0) = 0$, it follows that if x is close to the origin, the series $g_1^*, \ldots, g_n^*$ converge at $(|x_1|, \ldots, |x_n|)$, say to $t_1, \ldots, t_n$ respectively, and $t = (t_1, \ldots, t_n) \in D(f)$. We thus suppose that x verifies those conditions. Then $g_1, \ldots, g_n$ also converge at x, and we write $g(x) = (g_1(x), \ldots, g_n(x))$. As $|g_i(x)| \leq t_i$ (remember the definition of g_i^*) for $1 \leq i \leq n$, we conclude that $g(x) \in D(f)$ (Proposition 2.2). It remains to show that $f(g_1, \ldots, g_n)$ converges at x to $f(g(x))$.

To that end, for every integer $p \geq 0$ we consider the series
$$f_p = \sum_{|\nu|\leq p} a_\nu \mathbf{x}^\nu, \quad h_p = f_p(g_1, \ldots, g_n).$$

By 2.4 we have $h_p = f_p(g(x))$, and since f converges at $g(x)$ this implies
$$\lim_{p \to \infty} f_p(g(x)) = f(g(x)).$$

Thus we have to prove that $f(g_1, \ldots, g_n)$ converges at x to $\lim_{p\to\infty} h_p(x)$. By 2.5 a),b)
$$f(g_1, \ldots, g_n) - h_p = (f - f_p)(g_1, \ldots, g_n) = \sum_{|\nu|>p} a_\nu g_1^{\nu_1} \cdots g_n^{\nu_n}$$

for every p; we will denote this series by $\sum_\nu c_{p\nu}\mathbf{x}^\nu$. Also, let $\sum_\nu d_{p\nu}\mathbf{x}^\nu$ be the series obtained by substitution of $g_1^*, \ldots, g_n^*$ in $\sum_{|\nu|>p} |a_\nu|\mathbf{x}^\nu$. We claim that $\sum_\nu d_{p\nu}\mathbf{x}^\nu$ converges at x and
$$\sum_\nu d_{p\nu}|\mathbf{x}|^\nu \leq \sum_{|\nu|>p} |a_\nu t^\nu|.$$

Indeed, if $I \subset \mathbb{N}^n$ is finite and $q = \max\{|\nu| \mid \nu \in I\}$ it holds
$$\sum_{\nu \in I} d_{p\nu}|\mathbf{x}|^\nu \leq \sum_{p<|\nu|\leq q} |a_\nu| \left(\sum |b_{1\mu}x^\mu|\right)^{\nu_1} \cdots \left(\sum |b_{n\mu}x^\mu|\right)^{\nu_n} =$$
$$= \sum_{p<|\nu|\leq q} |a_\nu t^\nu| \leq \sum_{p<|\nu|} |a_\nu t^\nu|,$$

where the first inequality follows from 2.5 a),b). Now our claim is a consequence of Proposition 1.3, since the series $\sum |a_\nu t^\nu|$ converges.

On the other hand, it is clear that $|c_{p\nu}| \leq d_{p\nu}$, and so
$$|(f(g_1, \ldots, g_n) - h_p)(x)| \leq \sum_\nu d_{p\nu}|x^\nu| \leq \sum_{p<|\nu|} |a_\nu t^\nu|.$$

All this means that $f(g_1, \ldots, g_n)$ converges at x and that, the last inequality being valid for all p's, $f(g_1, \ldots, g_n)(x) = \lim_{p\to\infty} h_p(x)$, as wanted. □

2 Power Series

A first consequence of the preceding result is the following: if f is a convergent power series with $f(0) = a \neq 0$, then the formal power series

$$\frac{1}{f} = \frac{1}{a} \sum \left(1 - \frac{1}{a}f\right)^{\nu_1}$$

constructed above is convergent. Whence $\mathcal{O}_n$ is a local ring whose maximal ideal $\mathfrak{m}_n$ consists of the convergent power series with order ≥ 1. Again, this ideal is generated by the indeterminates: $\mathfrak{m}_n = \{x_1, \ldots, x_n\}\mathcal{O}_n$.

(2.7) Derivatives. Let $1 \leq i \leq n$. The *derivative with respect to* x_i of a formal power series $f = \sum a_\nu x^\nu$ is the formal power series

$$\partial f/\partial x_i = \frac{\partial f}{\partial x_i} = \sum_{\nu_i > 0} \nu_i a_\nu x_1^{\nu_1} \cdots x_i^{\nu_i - 1} \cdots x_n^{\nu_n}.$$

If $\{f_\lambda \mid \lambda \in \Lambda\}$ is a summable family of formal power series, the family $\{\partial f_\lambda/\partial x_i \mid \lambda \in \Lambda\}$ is also summable, and its sum is $\partial(\sum f_\lambda)/\partial x_i$. In the same way, the usual properties of derivatives hold true in this formal setting. The *Leibnitz Formula*:

$$\frac{\partial(fg)}{\partial x_i} = f\frac{\partial g}{\partial x_i} + g\frac{\partial f}{\partial x_i};$$

the *Chain Rule*:

$$\frac{\partial(f(g_1, \ldots, g_n))}{\partial x_i} = \sum_{1 \leq j \leq n} \frac{\partial f}{\partial x_i}(g_1, \ldots, g_n)\frac{\partial g_j}{\partial x_i};$$

the *Schwartz Rule*:

$$\frac{\partial}{\partial x_i}\left(\frac{\partial f}{\partial x_j}\right) = \frac{\partial}{\partial x_j}\left(\frac{\partial f}{\partial x_i}\right).$$

This last formula gives way to the definition by induction of the *derivatives of higher order*

$$\partial^{|\nu|}f/\partial x^\nu = \frac{\partial^{|\nu|}f}{\partial x^\nu}$$

where ∂x^ν stands for $\partial x_1^{\nu_1} \cdots \partial x_n^{\nu_n}$. In particular we get the *Taylor Expansion*:

$$f = \sum \frac{1}{\nu!}\frac{\partial^{|\nu|}f}{\partial x^\nu}(0)x^\nu.$$

More generally, for $\mathbf{x} = (x_1, \ldots, x_n), \mathbf{y} = (y_1, \ldots, y_p)$ and $f \in \mathbb{K}[[\mathbf{x}, \mathbf{y}]]$ we have

$$f(\mathbf{x}, \mathbf{y}) = \sum \frac{1}{\nu!}\frac{\partial^{|\nu|}f}{\partial \mathbf{x}^\nu}(0, \mathbf{y})\mathbf{x}^\nu.$$

Indeed, the family $\{\frac{1}{\nu!}\frac{\partial^{|\nu|}f}{\partial \mathbf{x}^\nu}(0, \mathbf{y})\mathbf{x}^\nu \mid \nu \in \mathbb{N}^n\}$ is summable, and if we denote by h its sum, a straightforward computation shows that $\omega(f - h) \geq m$ for any integer $m \geq 0$, that is, $f = h$.

For convergent power series we have:

Proposition 2.8 *Let f be a convergent power series. Then the associated function $^a f : D(f) \to \mathbb{K}$ is analytic, that is, $^a f$ is smooth and for every $x_0 \in D(f)$:*

a) The series $\dfrac{\partial^{|\nu|} f}{\partial x^\nu}$ converges at x_0 to $\dfrac{\partial^{|\nu|\,a} f}{\partial x^\nu}(x_0)$, and

b) The power series

$$T_{x_0} f = \sum \frac{1}{\nu!} \frac{\partial^{|\nu|} f}{\partial x^\nu}(x_0) x^\nu$$

is convergent, and for x close enough to x_0 it holds $f(x) = T_{x_0} f(x - x_0)$.

Proof. Let us see first that the partial derivative $\dfrac{\partial^a f}{\partial x_i}(x_0)$ exists and that $\dfrac{\partial f}{\partial x_i}$ converges at x_0 to that derivative. Without loss of generality assume $i = 1$. Using the notations of Proposition 2.3, we put $(1) = (1, 0, \ldots, 0)$, and so

$$g = b_{(1)} \mathbf{x}_1 + \mathbf{x}_1 g_1 + g_2; \quad g_1(0) = 0, \ g_2 = \sum_{\mu_1 = 0} \mathbf{x}^{\mu_1}.$$

Note that g_1 and g_2 are parts of the expansion of g, and consequently both are convergent. Now for small $t \neq 0$ it is

$$\frac{^a g(t, 0, \ldots, 0) - {^a g}(0)}{t} = b_{(1)} + {^a g_1}(t, 0, \ldots, 0),$$

and since $^a g_1$ is continuous and $g_1(0) = 0$, we deduce that the derivative $\dfrac{\partial^a g}{\partial x_1}(0)$ exists and

$$\frac{\partial^a g}{\partial x_1}(0) = b_{(1)} = \sum a_\nu P_{\nu,(1)}(x_0).$$

We also have

$$P_{\nu,(1)} = \nu_1 \mathbf{x}_1^{\nu_1 - 1} \mathbf{x}_2^{\nu_2} \cdots \mathbf{x}_n^{\nu_n},$$

whence

$$\sum a_\nu P_{\nu,(1)}(\mathbf{x}) = \frac{\partial f}{\partial x_1}$$

and we conclude that the latter series converges at x_0 to $b_{(1)}$. Thus

$$\frac{\partial f}{\partial x_1}(x_0) = \frac{\partial^a g}{\partial x_1}(0).$$

Finally by Proposition 2.3 *b)* we conclude that the derivative $\dfrac{\partial^a f}{\partial x_1}(x_0)$ exists and coincides with $\dfrac{\partial^a g}{\partial x_1}(0)$, that is

$$\frac{\partial^a f}{\partial x_1}(x_0) = \frac{\partial f}{\partial x_1}(x_0).$$

It follows easily by induction from this both that $^a f$ is smooth and the assertion *a)* in the statement. But then *b)* also follows, since the series $T_{x_0} f$ is exactly the series g above. □

Proposition 2.9 *(Identity Principle) Let f be a convergent power series. The following assertions are equivalent:*

a) $f = 0$.

b) $^a f$ *vanishes on a non-empty open subset of $D(f)$.*

c) *f and all its derivatives of all orders vanish at some point of $D(f)$.*

Proof. Let $A \subset D(f)$ be the set of all $x \in D(f)$ such that $\omega(T_x f) = +\infty$. By Proposition 2.8 the set A is both open and closed, and since $D(f)$ is connected (Proposition 2.2), A is either empty or equal to $D(f)$. Now if c) holds, $A \neq \emptyset$, and by the preceding remark $A = D(f)$, that is, $f = 0$. The other implications are trivial. □

Corollary 2.10 *Let $f \in \mathbb{K}[x_1, \ldots, x_n]$ be a non-zero polynomial. Then the open set $\{x \in \mathbb{K}^n \mid f(x) \neq 0\}$ is a dense subset of $\mathbb{K}^n$.*

3 Rückert's and Weierstrass's Theorems

A formal power series $f \in \mathcal{F}_n = \mathbb{K}[[x]]$, $x = (x_1, \ldots, x_n)$, is called *regular of order p with respect to x_n* if $f(0, \ldots, 0, x_n) = x_n^p g(x_n)$ with $g(0) \neq 0$. A polynomial $z^p + a_1 z^{p-1} + \cdots + a_p$ with coefficients $a_1, \ldots, a_p \in \mathcal{F}_n$ is called *distinguished* if it is a regular series of order p with respect to z, or equivalently if $a_1(0) = \cdots = a_p(0) = 0$.

The following lemma will be often useful:

Lemma 3.1 *Let $f \in \mathcal{F}_n$, $f \neq 0$. After a linear change of coordinates, f becomes regular of order $\omega(f)$ with respect to x_n.*

Proof. Set $f = \sum a_\nu x^\nu$, $p = \omega(f)$. Then $f_p = \sum_{|\nu|=p} a_\nu x^\nu \neq 0$, and there are $c_1, \ldots, c_{n-1} \in \mathbb{K}$ with $c = f_p(c_1, \ldots, c_{n-1}, 1) \neq 0$ (otherwise the homogeneous polynomial f_p would be divisible by $x_n - 1$). We now make the change of coordinates $x_i = y_i + c_i y_n, 1 \leq i < n$, $x_n = y_n$ to get $g(y) = f(y_1 + c_1 y_n, \ldots, y_{n-1} + c_{n-1} y_n, y_n)$. Clearly $g(0, \ldots, 0, y_n) = f(c_1 y_1, \ldots, c_{n-1} y_{n-1}, y_n)$ consists of the monomial $c y_n^p$ plus terms of higher degrees. □

Proposition 3.2 *(Rückert's Division Theorem) Let $\Phi \in \mathcal{O}_n$ a convergent power series, regular of order p with respect to x_n. For every $f \in \mathcal{O}_n$ there exist $Q \in \mathcal{O}_n$ and $R \in \mathcal{O}_{n-1}[x_n]$ with degree of $R < p$ such that $f = Q\Phi + R$. This conditions determine Q and R uniquely. Furthermore, if Φ is distinguished in x_n and $f \in \mathcal{O}_{n-1}[x_n]$, then $Q \in \mathcal{O}_{n-1}[x_n]$, too.*

The same result holds true when substituting $\mathcal{O}_n$ by $\mathcal{F}_n$ and $\mathcal{O}_{n-1}$ by $\mathcal{F}_{n-1}$.

Proof. Since Φ is regular of order p with respect to $\mathbf{x}_n$ we can write

$$\Phi = \varphi + c\mathbf{x}_n^p, \quad \varphi = \sum_{i=0}^{p} a_i(\mathbf{x}')\mathbf{x}_n^{p-i} + \mathbf{x}_n^{p+1} b(\mathbf{x}),$$

where $a_0, \ldots, a_p \in \mathcal{O}_{n-1} = \mathbb{K}[[\mathbf{x}']]$, $\mathbf{x}' = (\mathbf{x}_1, \ldots, \mathbf{x}_{n-1})$ and $c \in \mathbb{K}$. Up to multiplication by $1/c$ we may assume $c = 1$.

Let $\rho = (\rho_1, \ldots, \rho_n)$, $\rho_i > 0$. For $f = \sum a_\nu \mathbf{x}^\nu \in \mathcal{O}_n$ we will denote by $\|f\|$ the sum of the series $\sum |a_\nu| \rho^\nu$ when this sum exists and $\|f\| = +\infty$ otherwise. Let X be the set of all series $f \in \mathcal{O}_n$ with $\|f\| < +\infty$. Clearly, if ρ is small enough, X contains any prescribed finite collection of convergent power series; in particular $f, \varphi, b, a_i \in X$.

We define a map $T: X \to X$ as follows: if $Q \in X$ set

$$f - \varphi Q = R + \mathbf{x}_n^p T(Q),$$

where $R \in \mathcal{O}_{n-1}[\mathbf{x}_n]$, and the degree of R is $< p$. This map T is *contractive*, that is,

$$\mathrm{dist}(T(Q), T(Q')) < c\,\mathrm{dist}(Q, Q')$$

for any $Q, Q' \in X$, where $0 < c < 1$ (dist stands for the distance associated to the norm $\|\cdot\|$). Indeed, let

$$f - \varphi Q = R + \mathbf{x}_n^p T(Q), \quad f - \varphi Q' = R' + \mathbf{x}_n^p T(Q').$$

Then

$$\varphi(Q' - Q) = R - R' + \mathbf{x}_n^p (T(Q) - T(Q')).$$

Computing norms in this equality, and taking into account that $R - R'$ is a polynomial in $\mathcal{O}_{n-1}[\mathbf{x}_n]$ of degree $< p$, we get

$$\|\mathbf{x}_n^p (T(Q) - T(Q'))\| \leq \|R - R'\| + \|\mathbf{x}_n^p (T(Q) - T(Q'))\| =$$
$$= \|R - R' + \mathbf{x}_n^p (T(Q) - T(Q'))\| = \|\varphi(Q - Q')\| \leq \|\varphi\| \|Q - Q'\|.$$

But $\|\mathbf{x}_n^p (T(Q) - T(Q'))\| = \rho_n^p \|T(Q) - T(Q')\|$, and we obtain

$$\mathrm{dist}(T(Q), T(Q')) \leq \frac{\|\varphi\|}{\rho_n^p} \mathrm{dist}(Q, Q').$$

We have

$$\frac{\|\varphi\|}{\rho_n^p} \leq \frac{1}{\rho_n^p} \sum_{i=0}^{p} \|a_i\| \rho_n^{p-i} + \rho_n \|b\|.$$

Now, since $\|b\| < +\infty$, $\rho_n \|b\| < \frac{1}{4}$ for ρ_n small enough. Thus fixed ρ_n, for $\rho_1, \ldots, \rho_{n-1}$ small enough also the other sumand is $< \frac{1}{4}$, since $a_i(0) = 0$. All in all, we conclude

$$\mathrm{dist}(T(Q), T(Q')) < \frac{1}{2} \mathrm{dist}(Q, Q').$$

3 Rückert's and Weierstrass's Theorems

We assume for the moment the following

Claim. $(X, \|\cdot\|)$ *is a Banach algebra.*

Under this assumption we can apply the classical

(Fixed Point Theorem) Let X be a complete metric space and $T: X \to X$ a contractive map. Then T has a unique fixed point.

Consequently and up to the claim, our T has a unique fixed point, say $Q \in X$. It follows from the very definition of T that

$$f = Q(\varphi + \mathbf{x}_n^p) + R = Q\Phi + R.$$

Thus we have shown the existence of the division. And actually also its uniqueness: if $f = Q'\Phi + R'$ with $Q \neq Q' \in \mathcal{O}_n$, $R' \in \mathcal{O}_{n-1}[\mathbf{x}_n]$ with degree of $R' < p$, we can always choose ρ so that $Q', R' \in X$ and then Q' would be a second fixed point of T, which is impossible.

This settles the first part of the theorem, except for the claim, whose proof we postpone still a little. Suppose now that $f, \Phi \in \mathcal{O}_{n-1}[\mathbf{x}_n]$ and Φ is distinguished in $\mathbf{x}_n$. Then by division of f by the monic polynomial Φ in the ring of polynomials $\mathcal{O}_{n-1}[\mathbf{x}_n]$, we find $f = Q'\Phi + R'$ with $Q', R' \in \mathcal{O}_{n-1}[\mathbf{x}_n]$ and degree of $R' < p$. By the uniqueness already proved $Q = Q'$, $R = R'$.

We finally come to the

Proof of the claim: The key fact here is that $(X, \|\cdot\|)$ is complete. To see this we consider a Cauchy sequence $(g_s)_{s\geq 0}$ of X, where $g_s = \sum a_{s\nu}\mathbf{x}^\nu$ for $s \geq 0$. Then every sequence $(a_{s\nu})_{s\geq 0}$ is a Cauchy sequence of $\mathbb{K}$ with a limit $a_\nu \in \mathbb{K}$. It follows that $g = \sum a_\nu \mathbf{x}^\nu$ belongs to X, and $g = \lim_{s\to\infty} g_s$.

To prove this we need to show that for any given $\varepsilon > 0$ and large enough s

$$\sum_\nu |a_\nu - a_{s\nu}|\rho^\nu = \|g - g_s\| < \varepsilon$$

holds. Furthermore, since the series in this inequality is a series of non-negative real numbers, it is enough to see that

$$\sum_{\nu \in I} |a_\nu - a_{s\nu}|\rho^\nu \leq \varepsilon/2$$

for any finite set of indices I. We notice that since $(g_s)_{s\geq 0}$ is a Cauchy sequence, we have for r, s large

$$\sum_\nu |a_{r\nu} - a_{s\nu}|\rho^\nu < \varepsilon/2.$$

Consequently

$$\sum_{\nu \in I} |a_{r\nu} - a_{s\nu}|\rho^\nu < \varepsilon/2$$

and taking the limit in this finite sum as $r \to \infty$, we get

$$\sum_{\nu \in I} |a_\nu - a_{s\nu}| \rho^\nu \leq \varepsilon/2,$$

as wanted.

The proof in the formal case is similar. The Banach space used in this case is $X = \mathcal{F}_n$ with the norm $\|f\| = \exp^{-v(f)}$, where $v(f)$ stands for the greatest integer $m \geq 0$ such that $f \in \{x_1, \ldots, x_{n-1}\}^m \mathcal{F}_n$. □

Proposition 3.3 *(Weierstrass's Preparation Theorem)* Let $\Phi \in \mathcal{O}_n$ be regular of order p with respect to x_n. Then there exist a distinguished polynomial $P \in \mathcal{O}_{n-1}[x_n]$ of degree p and a unit Q of $\mathcal{O}_n$ such that $P = Q\Phi$. These conditions determine P and Q uniquely.

The same result holds true when substituting $\mathcal{O}_n$ by $\mathcal{F}_n$ and $\mathcal{O}_{n-1}$ by $\mathcal{F}_{n-1}$.

Proof. By the Division Theorem (Proposition 3.2) there exist $Q \in \mathcal{O}_n$, $a_1, \ldots, a_p \in \mathcal{O}_{n-1}$ such that

$$x_n^p = Q\Phi - \sum_{i=1}^{p} a_i x_n^{p-i}.$$

But $\Phi(0, x_n) = x_n^p g(x_n)$ with $g(0) \neq 0$, and consequently

$$x_n^p = Q(0, x_n) x_n^p g(x_n) - \sum_{i=1}^{p} a_i(0) x_n^{p-i}.$$

Thus we see that $a_1(0) = \cdots = a_p(0) = 0$, $Q(0,0) \neq 0$, and so

$$P = x_n^p + a_1 x_n^{p-1} + \cdots + a_p$$

is the distinguished polynomial we sought. The uniqueness of the division implies the uniqueness of P.

The proof in the formal case is the same. □

Proposition 3.4 *(Hensel's Lemma)* Let $\Phi \in \mathcal{O}_n[z]$ be a monic polynomial and $a \in \mathbb{K}$ a root of multiplicity p of $\Phi(0, z) \in \mathbb{K}[z]$. Then there exist monic polynomials $P, U \in \mathcal{O}_n[z]$ such that $\Phi = PU$ and

a) P has degree p and $P(0, z) = (z - a)^p$;

b) $U(0, a) \neq 0$.

These conditions determine P and U uniquely.

The same result holds true when substituting $\mathcal{O}_n$ by $\mathcal{F}_n$ and $\mathcal{O}_{n-1}$ by $\mathcal{F}_{n-1}$.

3 Rückert's and Weierstrass's Theorems

Proof. Up to the change $\mathbf{z}' = \mathbf{z} - a$ we may assume $a = 0$, and then our hypothesis implies that Φ is regular of order p with respect to $\mathbf{z}$. We thus can apply Weierstrass's Preparation Theorem (Proposition 3.3) to get P and $U = Q^{-1}$. We notice that *a)* means precisely that P is distinguished. Moreover, as P is distinguished and Φ a polynomial, U is a polynomial, too (last part of Rückert's Division Theorem). Finally the uniqueness follows from the uniqueness of Weierstrass's Preparation Theorem.

The same proof works in the formal case. □

II Analytic Rings and Formal Rings

Summary. We devote this chapter to the description of the categories of analytic and formal rings over $\mathbb{R}$ and $\mathbb{C}$. First, we develop Mather's formalism of finite and quasifinite homomorphisms. Then, we obtain Noether's Projection Lemma and further algebraic properties of these rings. Thus, we come to one fundamental construction: Abhyankar's and Rückert's Parametrization. Afterwards, we introduce the regularity ideals and prove Nagata's Jacobian Criteria. Finally, we discuss complexification, an essential tool to understand the differences between the real and the complex categories.

1 Mather's Preparation Theorem

Again we set $\mathbb{K} = \mathbb{R}$ or $\mathbb{C}$. Given a ring A and a prime ideal $\mathfrak{p}$ of A, we will denote by $\kappa(\mathfrak{p})$ the residue field of $\mathfrak{p}$, that is, the quotient field of the ring $A/\mathfrak{p}$; if A is local, we will denote by $\mathfrak{m}_A$ its maximal ideal.

Definition 1.1 *An* analytic *(resp. a* formal*) ring over $\mathbb{K}$ is a ring isomorphic to $\mathbb{K}\{\mathbf{x}\}/I$ (resp. $\mathbb{K}[[\mathbf{x}]]/I$) with $\mathbf{x} = (\mathbf{x}_1, \ldots, \mathbf{x}_n)$; we will usually not specify "over $\mathbb{K}$". If A, B are two analytic (resp. formal) rings, an* analytic *(resp. a* formal*) homomorphism $A \to B$ is a homomorphism of $\mathbb{K}$-algebras. The field $\mathbb{K}$ is called the* coefficient field*.*

Proposition 1.2 *An analytic (resp. a formal) ring is a noetherian local ring. The canonical homomorphism $\mathbb{K} \to A/\mathfrak{m}_A$ is an isomorphism, and $A = \mathbb{K} + \mathfrak{m}_A$.*

Proof. We will only give the proof in the analytic case, that of the formal case being analogous with the obvious changes. This will be done systematically all through this chapter.

First of all, since there is a certain surjective homomorphism $\mathcal{O}_n \to A$, the general case follows immediately from the case $A = \mathcal{O}_n$. Hence we suppose $A = \mathcal{O}_n$.

We argue by induction on n. If $n = 0$ the result is trivial, so we let $n > 0$ and I an ideal $\neq 0$ of $\mathcal{O}_n$. Choose $\Phi \in I$, $\Phi \neq 0$. By Lemma I.3.1, we may assume that Φ is regular of order, say, p with respect to $\mathbf{x}_n$. By Rückert's Division Theorem (Proposition I.3.2), the ring $\mathcal{O}_n/\Phi\mathcal{O}_n$ is generated by $1, \mathbf{x}_n, \ldots, \mathbf{x}_n^{p-1}$ as an $\mathcal{O}_{n-1}$-module. Since $\mathcal{O}_{n-1}$ is noetherian by induction hypothesis, we deduce that $\mathcal{O}_n/\Phi\mathcal{O}_n$ is a noetherian $\mathcal{O}_{n-1}$-module ([A-McD 6.5, 6.2]). Thus $I/\Phi\mathcal{O}_n$ is finitely generated as an $\mathcal{O}_{n-1}$-module, say by the classes of $f_1, \ldots, f_s \in I$. In this situation $f_1, \ldots, f_s, \Phi$ generate I.

Finally, the assertions concerning the coefficient field $\mathbb{K}$ are immediate. □

1 Mather's Preparation Theorem

Proposition 1.3 *Every analytic (resp. formal) homomorphism $\varphi : A \to B$ is local, that is, $\varphi(\mathfrak{m}_A)B \subset \mathfrak{m}_B$.*

Let $A = \mathcal{O}_n/I$ (resp. $\mathcal{F}_n/I$), and consider the correspondence

$$\Gamma : \varphi \mapsto (\varphi(\mathbf{x}_1 \bmod I), \ldots, \varphi(\mathbf{x}_n \bmod I)) \in \mathfrak{m}_B \times \cdots \times \mathfrak{m}_B = \mathfrak{m}_B^{\times n}.$$

from the set of all analytic (resp. formal) homomorphisms $A \to B$ into $\mathfrak{m}_B^{\times n}$. We have:

a) Γ is an injective map.

b) If $I = (0)$, then Γ is a bijective map.

Proof. First of all, if $\varphi(\mathfrak{m}_A)B \not\subset \mathfrak{m}_B$ there would be some $f \in \mathfrak{m}_A$ such that $\varphi(f) = a + g$ with $0 \neq a \in \mathbb{K}$, $g \in \mathfrak{m}_B$. Then $\varphi(f - a) = g$ would not be a unit, while $f - a$ is one.

a) We have to show that if $\phi, \varphi : A \to B$ are two analytic homomorphism such that $\phi(\mathbf{x}_i \bmod I) = \varphi(\mathbf{x}_i \bmod I) = b_i$, $1 \leq i \leq n$, then $\phi = \varphi$.

To see this, we note that any $f \in \mathcal{O}_n$ can be written in the form

$$f = g + h, \quad g \in \mathbb{K}[\mathbf{x}_1, \ldots, \mathbf{x}_n], \quad \omega(h) \geq s,$$

for every integer $s \geq 0$. Consequently

$$\phi(f \bmod I) - \varphi(f \bmod I) = \phi(h \bmod I) - \varphi(h \bmod I) \in \mathfrak{m}_B^s.$$

Indeed,

- ϕ and φ coincide on $\mathbf{x}_1 \bmod I, \ldots, \mathbf{x}_n \bmod I$, and both are $\mathbb{K}$-algebra homomorphisms; hence they coincide on $\mathbb{K}[\mathbf{x}_1 \bmod I, \ldots, \mathbf{x}_n \bmod I]$.

- $h \bmod I \in \mathfrak{m}_A^s$ and our homomorphisms are local.

This being valid for every $s \geq 0$, we may conclude

$$\phi(f \bmod I) - \varphi(f \bmod I) \in \bigcap_{s \geq 0} \mathfrak{m}_B^s = (0)$$

(Krull's Theorem [A-McD 10.19]).

b) Suppose now that $I = (0)$. To show that Γ is surjective fix any $b_1, \ldots, b_n \in \mathfrak{m}_B$. By the definition of an analytic ring, there is a surjective analytic homomorphism $\pi : \mathcal{O}_p \to B$, and we pick $g_1, \ldots, g_n \in \mathcal{O}_p$ such that $\pi(g_i) = b_i$, $1 \leq i \leq n$. We define $\varphi : A \to B$ by substitution:

$$f \mapsto f(g_1, \ldots, g_n) \mapsto \varphi(f) = \pi(f(g_1, \ldots, g_n)),$$

and so $\Gamma(\varphi) = (b_1, \ldots, b_n)$. $\square$

Definitions 1.4 *Let $\varphi : A \to B$ be a local homomorphism of local rings.*

a) φ is called quasifinite *if $B/\varphi(\mathfrak{m}_A)B$ has finite dimension as linear space over $\mathbb{K} = A/\mathfrak{m}_A$ (via φ).*

b) φ is called finite *if B is a finite A-module (via φ).*

Obviously every finite homomorphism is quasifinite. In our setting the converse is also true, as we see in the next result:

Proposition 1.5 *(Mather's Finiteness Theorem) Let $\varphi : A \to B$ be an analytic (resp. a formal) homomorphism. If M is a finite B-module and $M/\varphi(\mathfrak{m}_A)M$ has finite dimension as linear space over $\mathbb{K} = A/\mathfrak{m}_A$ (via φ), then M is a finite A-module (via φ).*

In particular, if φ is quasifinite, it is finite.

Proof. Choose two surjective analytic homomorphisms $\pi : \mathcal{O}_n \to A$, $\pi' : \mathcal{O}_p \to B$, where
$$\mathcal{O}_n = \mathbb{K}\{\mathbf{x}\},\ \mathcal{O}_p = \mathbb{K}\{\mathbf{y}\},\ \mathcal{O}_{n+p} = \mathbb{K}\{\mathbf{x},\mathbf{y}\},\ \mathbf{x} = (\mathbf{x}_1,\ldots,\mathbf{x}_n),\ \mathbf{y} = (\mathbf{y}_1,\ldots,\mathbf{y}_p).$$

We consider the inclusion $j : \mathcal{O}_n \to \mathcal{O}_{n+p}$ and power series $g_1,\ldots,g_n \in \mathcal{O}_p$ such that $\varphi\pi(\mathbf{x}_i) = \pi'(g_i)$ for $1 \le i \le n$. We then have the analytic homomorphism $\phi : \mathcal{O}_{n+p} \to \mathcal{O}_p$ defined by
$$\phi(\mathbf{x}_1) = g_1, \ldots, \phi(\mathbf{x}_n) = g_n, \phi(\mathbf{y}_1) = \mathbf{y}_1, \ldots, \phi(\mathbf{y}_p) = \mathbf{y}_p$$

(Proposition 1.3). We finally set $\pi'' = \pi' \circ \phi : \mathcal{O}_{n+p} \to B$, which is surjective. Clearly $\varphi\pi(\mathbf{x}_i) = \pi''(\mathbf{x}_i)$ for $1 \le i \le n$, and again by Proposition 1.3, we see that $\varphi \circ \pi = \pi''$. Hence we obtain the commutative diagram

$$\begin{array}{ccccc} \mathcal{O}_n & \xrightarrow{j} & \mathcal{O}_{n+p} & \xrightarrow{\phi} & \mathcal{O}_p \\ \downarrow{\pi} & & \downarrow{\pi''} & & \downarrow{\pi'} \\ A & & \xrightarrow{\varphi} & & B \end{array}$$

From this diagram one easily deduces that it suffices to prove the proposition for the homomorphism j, which by induction is reduced to the case $p = 1$.

Consequently we set $\mathbf{z} = \mathbf{y}_1$ and consider generators $m_1,\ldots,m_s$ of M, as $\mathcal{O}_{n+1}$-module, whose classes mod $\mathfrak{m}_n M$ generate $M/\mathfrak{m}_n M$ as linear space over $\mathbb{K} = \mathcal{O}_n/\mathfrak{m}_n$. Then
$$\mathbf{z} m_i = \sum_{j=1}^{s}(c_{ij} + h_{ij})m_j, \quad 1 \le i \le s,$$
for suitable $c_{ij} \in \mathbb{K}$, $h_{ij} \in \mathfrak{m}_n \mathcal{O}_{n+1}$. We now have the power series
$$\Phi = \det(\mathbf{z}\delta_{ij} - c_{ij} - h_{ij}) \in \mathcal{O}_{n+1}$$

1 Mather's Preparation Theorem

(where as usual $\delta_{ij} = 1$ if $i = j$, $\delta_{ij} = 0$ otherwise). But then

$$\Phi(0,\mathbf{z}) = \det(\mathbf{z}\delta_{ij} - c_{ij}) = \mathbf{z}^s + \sum_{k=1}^{s} c_k^* \mathbf{z}^{s-k} \in \mathbb{K}[\mathbf{z}],$$

since $h_{ij}(0, \mathbf{z}) = 0$. Hence $\Phi \in \mathcal{O}_{n+1}$ is regular of order $\leq s$ with respect to $\mathbf{z}$. On the other hand Φ is the determinant of the homogeneous system above, the m_i's seen as unknowns, and so we get

$$\Phi m_1 = \cdots = \Phi m_s = 0.$$

We finally consider $m \in M$. From Rückert's Division Theorem we get

$$m = \sum_{i=1}^{s} f_i m_i = \sum_{i=1}^{s}(Q_i \Phi + \sum_{j=1}^{s} u_{ij} \mathbf{z}^{p-j}) m_i = \sum_{1 \leq i,j \leq s} u_{ij} \mathbf{z}^{p-j} m_i,$$

with $f_i, Q_i \in \mathcal{O}_{n+1}$, $u_{ij} \in \mathcal{O}_n$. Thus the products $\mathbf{z}^{p-j} m_i$, $1 \leq i, j \leq s$, generate M as $\mathcal{O}_n$-module, which ends the proof. □

Example 1.6 Every non-trivial analytic homomorphism $\varphi : A \to \mathbb{K}\{\mathbf{t}\}$ (resp. formal homomorphism $\varphi : A \to \mathbb{K}[[\mathbf{t}]]$) is finite.
Proof. Indeed, $I = \varphi(\mathfrak{m}_A)\mathbb{K}\{\mathbf{t}\}$ is an ideal $\neq 0$. Then, if $0 \neq f \in I$ with $\omega(f) = d < +\infty$, we can write $f = \mathbf{t}^d u(\mathbf{t})$, where u is a unit. Thus $\mathbf{t}^d \in I$ and consequently I contains all power series of order $\geq d$. Hence, the homomorphism

$$\mathbb{K}^d \to \mathbb{K}\{\mathbf{t}\}/I : (a_0, \ldots, a_{d-1}) \mapsto a_0 + \cdots + a_{d-1} \mathbf{t}^{d-1} \mod I$$

is surjective. Consequently

$$\dim_{\mathbb{K}}(\mathbb{K}\{\mathbf{t}\}/I) \leq d,$$

and by Mather's Finiteness Theorem, φ is finite. □

This example is a very particular instance of a useful finiteness criterion that follows from Mather's:

Proposition 1.7 *Let $\varphi : A \to B$ be an analytic (resp. a formal) homomorphism. The following assertions are equivalent:*

a) *φ is finite.*

b) *$\mathfrak{m}_B = \sqrt{\varphi(\mathfrak{m}_A)B}$.*

Proof. Set $I = \varphi(\mathfrak{m}_A)B$ and $N_s = \mathfrak{m}_B^s + I$ for $s \geq 0$; clearly $N_s \supset N_{s+1}$. Then, by means of any surjective analytic homomorphism $\phi : \mathcal{O}_p \to B$, we obtain a surjective homomorphism

$$\mathfrak{m}_p^s / \mathfrak{m}_p^{s+1} \to N_s / N_{s+1}.$$

Now note that $\mathfrak{m}_p^s/\mathfrak{m}_p^{s+1}$ is the space of all homogeneous linear forms of degree exactly s, which has finite dimension as vector space over $\mathbb{K}$. Hence we also have

$$\dim_{\mathbb{K}}(N_s/N_{s+1}) < +\infty.$$

After this preparation, suppose φ finite. Then $\dim_{\mathbb{K}}(B/I) < +\infty$ and the chain

$$B = N_0 \supset N_1 \supset \cdots \supset N_s \supset N_{s+1} \supset \cdots \supset I$$

must be finite, that is, $N_s = N_{s+1}$ for some s. It follows $N_s = \mathfrak{m}_B N_s + I$ and applying Nakayama's Lemma ([A-McD 2.7]) to N_s as B-module we conclude

$$I = N_s = \mathfrak{m}_B^s + I \supset \mathfrak{m}_B^s.$$

As $\mathfrak{m}_B$ is the maximal ideal of B the above inclusion means that $\mathfrak{m}_B = \sqrt{I}$, as wanted.

Conversely, suppose b). Then, since the ring B is noetherian, $I \supset \mathfrak{m}_B^s$ for some s. Hence

$$\dim_{\mathbb{K}}(B/I) \leq \sum_{\ell=0}^{s} \dim_{\mathbb{K}}(N_\ell/N_{\ell+1}) < +\infty,$$

and φ is quasifinite. By Mather's Finiteness Theorem, φ is finite. □

Another consequence of Proposition 1.5 is:

Proposition 1.8 *Let $A = \mathcal{O}_n$ (resp $\mathcal{F}_n$) and $\varphi : A \to B$ be an analytic (resp. a formal) homomorphism with $\varphi(\mathbf{x}_i) = b_i \in \mathfrak{m}_B$ for $1 \leq i \leq n$. Then φ is surjective if and only if the b_i's generate $\mathfrak{m}_B$.*

Proof. The only if part is clear, since $\varphi^{-1}(\mathfrak{m}_B) \subset \mathfrak{m}_A$ and

$$\varphi(\mathfrak{m}_A)B = \{\varphi(\mathbf{x}_1), \ldots, \varphi(\mathbf{x}_n)\}B = \{b_1, \ldots, b_n\}B.$$

Conversely, if the b_i's generate $\mathfrak{m}_B$, then $\varphi(\mathfrak{m}_A)B = \mathfrak{m}_B$. Hence φ is quasifinite and, by Mather's theorem, it is finite. Thus, we can apply Nakayama's Lemma to B as a finite A-module, and since

$$\varphi(A) + \varphi(\mathfrak{m}_A)B \supset \mathbb{K} + \mathfrak{m}_B = B$$

we conclude $\varphi(A) = B$, and φ is onto. □

Finally, dimension theory can be applied to analytic and formal rings (for dimension theory we refer to [A-McD §11]). The following fact will we often used:

Lemma 1.9 *An analytic (resp. a formal) ring is regular of dimension n if and only if it is isomorphic to $\mathcal{O}_n$ (resp. $\mathcal{F}_n$).*

Proof. Firstly, $\mathcal{O}_n$ has dimension $\geq n$, since we have the chain
$$(0) \subset (\mathbf{x}_1) \subset \cdots \subset (\mathbf{x}_1, \ldots, \mathbf{x}_n) = \mathfrak{m}_n.$$
On the other hand, the ideal $\mathfrak{m}_n$ is generated by the n elements $\mathbf{x}_1, \ldots, \mathbf{x}_n$, and this implies that $\mathcal{O}_n$ is regular of dimension exactly n ([A-McD 11.14, 11.22]).

Now, let B be an analytic ring of dimension n. If B is regular, $\mathfrak{m}_B$ is generated by n elements $b_1, \ldots, b_n$, which define an analytic homomorphism $\varphi : \mathcal{O}_n \to B$. By Proposition 1.8 this homomorphism is onto, and it remains to see that it is injective. To that end, we put $I = \ker(\varphi)$ and consider the isomorphism $\mathcal{O}_n/I \simeq B$ induced by φ. We get $\dim(\mathcal{O}_n/I) = \dim(B) = n$, which clearly implies $I = (0)$. Whence, we are done. □

2 Noether's Projection Lemma

We start this section by studying factorization in power series rings.

Proposition 2.1 *The ring $\mathcal{O}_n$ (resp. $\mathcal{F}_n$) is factorial and, consequently, integrally closed in its quotient field.*

Proof. Since the ring $\mathcal{O}_n$ is a noetherian integral domain, it is enough to see that any irreducible element $\Phi \in \mathcal{O}_n$ generates a prime ideal $\Phi\mathcal{O}_n$. We will do this by induction. For $n = 0$ the assertion is trivial, so we assume $n > 0$ and the result to be known for less than n indeterminates. After a linear change of coordinates we may suppose that Φ is regular of some order with respect to $\mathbf{x}_n$ (Lemma I.3.1) and then, by Weierstrass's Preparation Theorem (Proposition I.3.3), that Φ is a distinguished polynomial of $\mathcal{O}_{n-1}[\mathbf{x}_n]$. By induction hypothesis the ring $\mathcal{O}_{n-1}$ is factorial, and by Gauss's Lemma ([L V.6 Th.10]), the ring $\mathcal{O}_{n-1}[\mathbf{x}_n]$ is also factorial.

Let us see now that Φ is irreducible in $\mathcal{O}_{n-1}[\mathbf{x}_n]$. Suppose $\Phi = \Phi_1 \Phi_2$ for some $\Phi_1, \Phi_2 \in \mathcal{O}_{n-1}$. Since Φ is irreducible in $\mathcal{O}_n$, one of those factors, say Φ_1 is a unit in $\mathcal{O}_n$, and $\Phi_2 = (1/\Phi_1)\Phi$ in $\mathcal{O}_n$. Then by Rückert's Division Theorem (Proposition I.3.2), since Φ is a distinguished polynomial and Φ_2 a polynomial, the quotient $1/\Phi_1$ must be a polynomial, too. Thus Φ_1 is a unit of $\mathcal{O}_{n-1}[\mathbf{x}_n]$, and we conclude that Φ is irreducible in $\mathcal{O}_{n-1}[\mathbf{x}_n]$.

We finally show that $\Phi\mathcal{O}_n$ is a prime ideal. Let Φ divide the product of two power series $f, g \in \mathcal{O}_n$. By Rückert's Division Theorem we can write
$$f = Q\Phi + R, \quad g = Q'\Phi + R'.$$
It follows $RR' = Q^*\Phi$ for some $Q^* \in \mathcal{O}_n$. Arguing as above, we now also conclude $Q^* \in \mathcal{O}_{n-1}[\mathbf{x}_n]$, and Φ being irreducible in the factorial ring $\mathcal{O}_{n-1}$, Φ divides either R or R'. Hence Φ divides either f or g.

Finally, the ring $\mathcal{O}_n$ is integrally closed in its quotient field. Actually, any factorial domain has this property, and the proof is the elementary one given for the ring of integers $\mathbb{Z}$. □

Remark 2.2 The preceding proof further shows that any distinguished polynomial $\Phi \in \mathcal{O}_{n-1}[\mathbf{x}_n]$ has a unique factorization $\Phi = P_1^{\alpha_1} \cdots P_s^{\alpha_s}$, where the P_i's are distinguished polynomials, irreducible both in $\mathcal{O}_{n-1}[\mathbf{x}_n]$ and $\mathcal{O}_n$.

We recall that the *height* $\mathrm{ht}(\mathfrak{p})$ of a prime ideal $\mathfrak{p}$ of a ring A is the maximal length of a chain of distinct prime ideals contained in $\mathfrak{p}$, that is, $\mathrm{ht}(\mathfrak{p}) = \dim(A_\mathfrak{p})$, and this height is finite if A is noetherian. For an arbitrary ideal I the height $\mathrm{ht}(I)$ is defined to be the minimal height of a prime ideal containing I. If A is noetherian, then $\sqrt{I}$ is a finite intersection of prime ideals: $\sqrt{I} = \mathfrak{p}_1 \cap \ldots \cap \mathfrak{p}_r$, and we get

$$\mathrm{ht}(I) = \mathrm{ht}(\sqrt{I}) = \min\{\mathrm{ht}(\mathfrak{p}_1), \ldots, \mathrm{ht}(\mathfrak{p}_r)\}.$$

Note also that if $\mathfrak{p} \subset I$ and $\mathrm{ht}(\mathfrak{p}) = \mathrm{ht}(I)$, then $\mathfrak{p} = I$.

(2.3) Transversal changes of coordinates. We describe here a construction that works the same for formal and convergent power series; as usual, we only discuss the convergent case.

> *a) Let I be an ideal of $\mathcal{O}_n$. After a linear change of coordinates there are distinguished polynomials $P_i \in I \cap \mathcal{O}_{n-i}[\mathbf{x}_{n-i+1}]$, $1 \leq i \leq r$, whose degrees coincide with their orders as power series, and such that $I \cap \mathcal{O}_{n-r} = (0)$.*

We will construct the linear change by induction. First we pick $P_1 \in I \setminus (0)$, and after a linear change as in Lemma I.3.1 and an application of Weierstrass's Preparation Theorem (Proposition I.3.3) this P_1 is the first polynomial we seek in $I \cap \mathcal{O}_{n-1}[\mathbf{x}_n]$. Then, having obtained P_i for $1 \leq i < j$, if $I \cap \mathcal{O}_{n-j+1} \neq (0)$, we choose $P_j \in I \cap \mathcal{O}_{n-j+1} \setminus (0)$. We again use Lemma I.3.1 to make P_j regular with respect to $\mathbf{x}_{n-j+1}$, but note that the linear change used only involves the indeterminates $\mathbf{x}_1, \ldots, \mathbf{x}_{n-j+1}$, and consequently does not affect the property that the already constructed polynomials are distinguished with respect to the other variables. We then apply Weierstrass's Preparation Theorem and may assume P_j is a distinguished polynomial, too. In that way, after say r steps we are done. □

> *b) After the linear change above, $\mathcal{O}_n/I$ is a finite $\mathcal{O}_{n-r}$-module.*

Indeed, let p_i denote the degree of P_i for $1 \leq i \leq r$. Any $f \in \mathcal{O}_n$ can be succesively divided by $P_1, \ldots, P_r$ as follows

$$\begin{aligned} f &= QP_1 + \sum_{k=1}^{p_1} a_k \mathbf{x}_n^{p_1-k} = \\ &= QP_1 + \sum_{k=1}^{p_1} \left(Q_k P_2 + \sum_{\ell=1}^{p_2} b_{k\ell} \mathbf{x}_{n-1}^{p_2-\ell} \right) \mathbf{x}_n^{p_1-k} = \end{aligned}$$

2 Noether's Projection Lemma

$$= H_1 P_1 + H_2 P_2 + \sum_{k,\ell} b_{k\ell} x_{n-1}^{p_2-\ell} x_n^{p_1-k} = \ldots$$

and in the end f will be written as an element $\sum_{i=1}^{r} H_i P_i \in I$ plus something generated by the monomials

$$x_{n-r+1}^{\nu_{n-r+1}} \cdots x_n^{\nu_n}, \quad 0 \leq \nu_{n-i+1} < p_i.$$

This proves b). □

c) We have $\mathrm{ht}(I) = r$.

First of all we deduce from b) that

$$\dim(\mathcal{O}_n/I) \geq \dim(\mathcal{O}_{n-r}) = n - r,$$

and consequently

$$\mathrm{ht}(I) \leq \dim(\mathcal{O}_n) - \dim(\mathcal{O}_n/I) \leq r.$$

Now, to get the opposite inequality, it is enough to show that

$$\mathrm{ht}(\{P_1, \ldots, P_r\} \mathcal{O}_n) \geq r,$$

and for this it suffices to see that

$$\mathrm{ht}(\{P_j, \ldots, P_r\} \mathcal{O}_n) > \mathrm{ht}(\{P_{j-1}, \ldots, P_r\} \mathcal{O}_n)$$

for $1 \leq j < r$ (note that $\mathrm{ht}(P_r \mathcal{O}_n) > 0$ since $P_r \neq 0$). To make notations clearer we will prove the case $j = 1$, that is, we will see that

$$\mathrm{ht}(\{P_1, \ldots, P_r\} \mathcal{O}_n) > \mathrm{ht}(\{P_2, \ldots, P_r\} \mathcal{O}_n).$$

Consider the the ideal

$$J = \sqrt{\{P_2, \ldots, P_r\} \mathcal{O}_{n-1}}.$$

Since $\mathcal{O}_{n-1}$ is noetherian, we have a decomposition

$$J = \mathfrak{p}_1 \cap \cdots \cap \mathfrak{p}_s,$$

where the $\mathfrak{p}_i$'s are prime ideals of $\mathcal{O}_{n-1}$. Now, every $f \in \mathcal{O}_n$ can be uniquely written in the form $\sum_{k \geq 0} a_k x_n^k$ with $a_k \in \mathcal{O}_{n-1}$, and using Krull's Theorem ([A-McD 10.19]) one sees that

$$\mathfrak{q}_i = \mathfrak{p}_i \mathcal{O}_n = \mathfrak{p}_i[[x_n]];$$

that is, an element f as above is in $\mathfrak{q}_i$ if and only if all the a_k's are in $\mathfrak{p}_i$. Using this description one easily deduces that $\mathfrak{q}_1, \ldots, \mathfrak{q}_s$ are prime ideals and again using Krull's Theorem that

$$\mathfrak{q}_1 \cap \cdots \cap \mathfrak{q}_s = \sqrt{\{P_2, \ldots, P_r\} \mathcal{O}_n}.$$

We suppose now that
$$\mathrm{ht}(\{P_1,\ldots,P_r\}\mathcal{O}_n) = \mathrm{ht}(\{P_2,\ldots,P_r\}\mathcal{O}_n).$$
Then both ideals share some associated prime among the $\mathfrak{q}_i$'s, say $\mathfrak{q}_1$, and in particular $P_1 \in \mathfrak{q}_1 = \mathfrak{p}_1[[\mathbf{x}_n]]$. This is however impossible: P_1 being monic of degree p_1, the coefficient 1 of $\mathbf{x}_n^{p_1}$ does not belong to $\mathfrak{p}_1$. □

We have actually proved that $\mathrm{ht}(I) + \dim(\mathcal{O}_n/I) = \dim(\mathcal{O}_n)$, a formula that is valid in a more general situation:

Proposition 2.4 *Let A be an analytic (resp. a formal) ring which is a domain, and $\mathfrak{p}$ a prime ideal of A. Then*
$$\mathrm{ht}(\mathfrak{p}) + \dim(A/\mathfrak{p}) = \dim(A).$$

Proof. We have (up to isomorphism) $A = \mathcal{O}_n/I$, and by the preceding construction we get a finite homomorphism $\mathcal{O}_d \hookrightarrow A$, where $d = \dim(\mathcal{O}_n/I)$. By the general properties of integral dependence ([A-McD 5.9, 5.16]):
$$\mathrm{ht}(\mathfrak{p}) = \mathrm{ht}(\mathfrak{p} \cap \mathcal{O}_d); \quad \dim(A/\mathfrak{p}) = \dim(\mathcal{O}_d/\mathfrak{p} \cap \mathcal{O}_d),$$
and since the formula of the statement holds for $\mathcal{O}_d$, it also holds for A. □

Corollary 2.5 *Let $\mathfrak{p} \supset \mathfrak{p}'$ be two prime ideals of an analytic (resp. a formal) ring A. Then all maximal chains of prime ideals in between $\mathfrak{p}$ and $\mathfrak{p}'$ have the same length.*

Proof. Apply Proposition 2.3 to the ideal $\mathfrak{p}/\mathfrak{p}'$ of the domain $A/\mathfrak{p}'$. □

We thus come to the main result of the section.

Proposition 2.6 *(Noether's Projection Lemma) Let $I \neq (0)$ be an ideal of heigth r of $\mathcal{O}_n$. Then:*

a) *After a linear change of coordinates the canonical homomorphism $\mathcal{O}_{n-r} = A \to B = \mathcal{O}_n/I$ is finite and injective.*

b) *Setting $\theta_j = \mathbf{x}_j + I$ for $n - r < j \leq n$, we have $B = A[\theta_{n-r+1},\ldots,\theta_n]$.*

c) *Suppose that I is prime and let K (resp. L) denote the quotient field of A (resp. B). After an additional linear change of coordinates involving only the last r indeterminates $\mathbf{x}_{n-r+1},\ldots,\mathbf{x}_n$, the element θ_{n-r+1} is a primitive element of L over K, that is, $L = K[\theta_{n-r+1}]$.*

All these linear changes can be done simultaneously for any given finite family of ideals of the same height.

The same result holds true in the formal case, that is, when replacing $\mathcal{O}_n$ by $\mathcal{F}_n$ and $\mathcal{O}_{n-r}$ by $\mathcal{F}_{n-r}$.

2 Noether's Projection Lemma

Proof. The first two statements follow from the construction 5.2, which can be applied to several ideals at a time. Now, if I is prime, we use the

(Primitive Element Theorem [L VII.6]) Let K be a field of characteristic zero and $C \subset K$ an infinite subset. If u, v are algebraic over K, there is $c \in C$ such that $K(u, v) = K(u + cv)$.

By induction we find $c_{n-r+1} = 1, \ldots, c_n \in \mathbb{K}$ such that

$$\sum_{k=n-r+1}^{n} c_k \mathbf{x}_k \bmod I$$

is a primitive element of L over K. We thus finish with the linear change

$$\mathbf{y}_i = \mathbf{x}_i, \text{ for } i \neq n-r+1; \quad \mathbf{y}_{n-r+1} = \sum_{k=n-r+1}^{n} c_i \mathbf{x}_k.$$

Finally, we claim also that this last change of coordinates can be done simultaneously for several ideals. Indeed, by the formulation of the Primitive Element Theorem chosen above, the set

$$C' = \{c \in C \mid K(u, v) = K(u + cv)\}$$

is infinite. Consequently, in the application of the theorem to any other field K' containing C and two elements u', v' algebraic over K', there is $c' \in C'$ such that $K'(u', v') = K'(u' + c'v')$. From this remark and by induction, our claims follow clearly. □

Remarks 2.7 *a)* A by-product of the preceding constructions is the following characterization of the dimension d of an analytic ring $\mathcal{O}_n/I$: d is the smallest number of independent homogeneous linear forms $h_1, \ldots, h_d$ such that

$$\sqrt{(h_1, \ldots, h_d) + I} = (\mathbf{x}_1, \ldots, \mathbf{x}_n).$$

For, suppose such linear forms are given. Then by Proposition 1.7 the homomorphism

$$\mathcal{O}_d \to \mathcal{O}_n/I : \mathbf{x}_1, \ldots, \mathbf{x}_d \mapsto h_1, \ldots, h_d$$

is finite. Hence $\dim(\mathcal{O}_n/I) \leq d$. Conversely, if d is the dimension of $\mathcal{O}_n/I$, after a transversal change of coordinates (2.3) the canonical homomorphism $\mathcal{O}_d \to \mathcal{O}_n/I$ is finite, and by Proposition 1.7 we would be done again. We can also argue directly as follows. By 2.3 *a)*, there are distinguished polynomials $P_i \in \mathcal{O}_{n-i}[t]$, $1 \leq i \leq r = n - d$, such that $P_i(\mathbf{x}_{n-i+1}) \in I$. Using these P_i's, we will see that $\mathbf{x}_{n-i+1} \in \sqrt{(\mathbf{x}_1, \ldots, \mathbf{x}_d) + I}$. For $i = r$ all coefficients of P_r belong to $(\mathbf{x}_1, \ldots, \mathbf{x}_d)$, and consequently the highest degree monomial of $P_r(\mathbf{x}_{d+1})$ belongs to $(\mathbf{x}_1, \ldots, \mathbf{x}_d) + I$. This highest degree monomial is a power of $\mathbf{x}_{d+1}$, and so $\mathbf{x}_{d+1} \in \sqrt{(\mathbf{x}_1, \ldots, \mathbf{x}_d) + I}$. Repeating this, we end by descending induction on i. □

b) The linear changes in all this section are *generic*, in the sense that they can be chosen arbitrarily close to the identity, and consequently to any other fixed linear change.

Indeed, the linear changes used have coefficients arbitrarily small off the diagonal, and this consists of 1's. If we are given a linear change other than the identity, we make it first, and then make another one close to the identity. □

3 Abhyankar's and Rückert's Parametrization

Let B be an analytic (resp. a formal) domain of dimension d, $A = \mathbb{K}\{\mathbf{x}\}$ (resp. $\mathbb{K}[[\mathbf{x}]]$), $\mathbf{x} = (x_1, \ldots, x_d)$, and $A \to B$ an analytic (resp. a formal) homomorphism. Suppose that $A \to B$ is finite and injective, so that the quotient field L of B is an extension of the quotient field K of A. Let $\theta \in \mathfrak{m}_B$. Then:

Lemma 3.1 *The irreducible polynomial P of θ over K has coefficients in A and is a distinguished polynomial $P(\mathbf{x}, \mathbf{t}) \in A[\mathbf{t}]$.*

Proof. Set $P = \mathbf{t}^p + a_1 \mathbf{t}^{p-1} + \cdots + a_p$, $a_i \in K$. Consider a field $F \supset L$ where P has its p roots $\theta = t_1, \ldots, t_p$ (all different since the characteristic is zero). We may also assume that there are K-automorphisms of F, $\sigma_1 = \mathrm{Id}_F, \ldots, \sigma_p$, such that $\sigma_i(\theta) = t_i$ for $1 \leq i \leq p$ ([L VII.2]). Since the homomorphism $A \to B$ is finite, θ is integral over A, that is, θ satisfies a monic equation with coefficients in $A \subset K$. Applying σ_i to such an equation we see that t_i satisfies also the resulting equation, and consequently t_i is integral over A. Finally, the t_i's are the roots of P, and so the coefficients a_j of P are symmetric functions of the t_i's. Thus the a_i's are also integral over A. But $a_1, \ldots, a_p \in K$ and A is integrally closed in K (Proposition 2.1), so that $a_1, \ldots, a_p \in A$. Whence $P \in A[\mathbf{t}]$.

Now note that P is clearly regular with respect to $\mathbf{t}$ of some order $s \leq p$ and consequently by Weierstrass's Preparation Theorem

$$P = P_1 U,$$

where $P_1 \in A[\mathbf{t}]$ is a distinguished polynomial of degree s and $U(0,0) \neq 0$. Moreover, by Rückert's Division Theorem, U is also a polynomial of $A[\mathbf{t}]$, and both P_1 and U are monic. It follows $P_1(\mathbf{x}, \theta) U(\mathbf{x}, \theta) = 0$, and since B is a domain, either $P_1(\mathbf{x}, \theta) = 0$ or $U(\mathbf{x}, \theta) = 0$. Hence, P being the irreducible polynomial of θ, we get either $P = P_1$ or $P = U$. But $\theta \in \mathfrak{m}_B$ and $\mathfrak{m}_B \cap A = \mathfrak{m}_A$, so that from

$$\theta^p + a_1 \theta^{p-1} + \cdots + a_p = 0$$

it follows $a_p \in \mathfrak{m}_A$. If it were $P = U$, we would have

$$U(\mathbf{x}, 0) = a_p \in \mathfrak{m}_A,$$

and so

… 3 Abhyankar's and Rückert's Parametrization

$$U(0,0) = a_p(0) = 0,$$

against the definition of U. Thus $P = P_1$ and P is distinguished. □

We next suppose that θ is a primitive element, $L = K[\theta]$, and we consider the discriminant $\delta \in A$ of P, that is, the resultant of P and its derivative $\partial P/\partial t$ ([L V.10]). Since P has no multiple roots, $\delta \neq 0$. We will denote by $\overline{B}$ the integral closure of B in L. In this situation:

Proposition 3.2 *We have $\delta \overline{B} \subset A + A\theta + \cdots + A\theta^{p-1}$, where p is the dimension $[L : K]$ of L as a K-linear space.*

Proof. We use the same notation as in the previous proof. Let $y \in \overline{B}$. Since θ is a primitive element we have

$$y = b_0 + b_1\theta + \cdots + b_{p-1}\theta^{p-1},$$

where the b_i's are elements of K. Applying the K-automorphisms σ_j to the equation above we obtain new equations

$$y_j \sigma_j(y) = b_0 + b_1 t_j + \cdots + b_{p-1} t_j^{p-1}, \quad 1 \leq j \leq p,$$

which we look at as a system on the b_i's. The determinant of this system is the Vandermonde determinant of the elements $t_1, \ldots, t_p$, which is well known to be

$$c = \prod_{1 \leq i < j \leq p} (t_j - t_i).$$

In particular $\delta = c^2$, and from Cramer's Rule we obtain

$$\delta b_i = c D_i(y_1, \ldots, y_p, t_1, \ldots, t_p),$$

where D_i is a polynomial with coefficients in $\mathbb{Z}$. We now recall that the t_j's are integral over A as θ is, and note that analogously the y_j's are integral over B as y is. But since $A \to B$ is finite, B is integral over A ([A-McD 5.1]), and the conclusion is that all those elements are integral over A. Whence every $\delta b_i \in K$ is integral over A and, since A is integrally closed in K, $\delta b_i \in A$. Thus we get

$$\delta y = \delta b_0 + \delta b_1 \theta + \cdots + \delta b_{p-1} \theta^{p-1} \in A + A\theta + \cdots + A\theta^{p-1}.$$

□

Corollary 3.3 *The ring $\overline{B}$ is a finitely generated B-module.*

Proof. In fact, $\delta \overline{B}$ is a sub A-module of $M = A + A\theta + \cdots + A\theta^{p-1}$, which is finitely generated over A. Since A is noetherian, M is noetherian, too, and $\delta \overline{B}$ is finitely generated over A. Finally, if $\delta g_1, \ldots, \delta g_m \in \delta \overline{B}$ generate $\delta \overline{B}$ over A, then $g_1, \ldots, g_m$ generate $\overline{B}$ over A, and, consequently, over B. □

Now we come back to the situation of Noether's Projection Lemma:

Proposition 3.4 *(Rückert's Parametrization) Let $\mathfrak{p}$ a prime ideal of $\mathcal{O}_n$ of height r, and put $d = n - r$. After a linear change of coordinates the following conditions hold:*

a) *The canonical homomorphism $\mathcal{O}_d = A \to B = \mathcal{O}_n/\mathfrak{p}$ is injective and finite.*

b) *The class $\theta_{d+1} = \mathbf{x}_{d+1} \bmod \mathfrak{p}$ is a primitive element of the quotient field L of B over the quotient field K of A.*

c) *The irreducible polynomial over K of $\theta_j = \mathbf{x}_j \bmod \mathfrak{p} \in \mathfrak{m}_B$ is a distinguished polynomial $P_j \in \mathcal{O}_d[\mathbf{x}_j]$ $(d < j \leq n)$.*

d) *For every $j = d+2, \ldots, n$, there is a polynomial $Q_j \in \mathcal{O}_d[\mathbf{x}_{d+1}]$ of degree $< p = $ degree of P_{d+1} such that*

$$\delta \mathbf{x}_j - Q_j \in \mathfrak{p},$$

where $\delta \in \mathcal{O}_d \setminus (0)$ is the discriminant of P_{d+1}.

e) *There is an integer $q \geq 1$ such that*

$$\delta^q \mathfrak{p} \subset I = \{P_{d+1}, \delta \mathbf{x}_{d+2} - Q_{d+2}, \ldots, \delta \mathbf{x}_n - Q_n\} \mathcal{O}_n \subset \mathfrak{p}.$$

As usual, there is an analogous statement in the formal case.

Proof. After a linear change of coordinates, conditions a)-d) follow from Proposition 2.5, Lemma 3.1 and Proposition 3.2. We will now prove that $\delta^q \mathfrak{p} \subset I$ for a suitable q. To do that, since $\mathfrak{p}$ is a finitely generated ideal, it suffices to see that for every $f \in \mathfrak{p}$ there is q such that $\delta^q f \in I$. Hence, we start by dividing f succesively by $P_{d+2}, \ldots, P_n$ until we get

$$f = \sum a_\nu \mathbf{x}_{d+2}^{\nu_{d+2}} \cdots \mathbf{x}_n^{\nu_n} \mod P_{d+2}, \ldots, P_n,$$

where the sum is finite and the a_ν's belong to $\mathcal{O}_{d+1}$ (see the proof of 2.3 b)). For a suitable $q \geq 1$ we can write

$$\delta^q f = \sum \delta^{q\nu} a_\nu (\delta \mathbf{x}_{d+2})^{\nu_{d+2}} \cdots (\delta \mathbf{x}_n)^{\nu_n} \mod P_{d+2}, \ldots, P_n. \tag{1}$$

On the other hand, put $\mathbf{x}' = (\mathbf{x}_1, \ldots, \mathbf{x}_d)$, and note that

$$P_j^*(\mathbf{x}', \mathbf{x}_{d+1}) = \delta^s P_j \left(\mathbf{x}', \frac{Q_j(\mathbf{x}', \mathbf{x}_{d+1})}{\delta} \right) \in \mathcal{O}_d[\mathbf{x}_{d+1}]$$

for large s. Next, since $\delta \mathbf{x}_j - Q_j \in \mathfrak{p}$, it follows

$$P_j^*(\mathbf{x}', \theta_{d+1}) = 0.$$

Now, as the irreducible polynomial of θ_{d+1} is P_{d+1},

$$P_j^*(\mathbf{x}', \mathbf{x}_{d+1}) = G(\mathbf{x}', \mathbf{x}_{d+1}) P_{d+1}(\mathbf{x}', \mathbf{x}_{d+1})$$

3 Abhyankar's and Rückert's Parametrization

where $G \in \mathcal{O}_d[\mathbf{x}_{d+1}]$. Summing up

$$\delta^s P_j\left(\mathbf{x}', \frac{Q_j(\mathbf{x}', \mathbf{x}_{d+1})}{\delta}\right) = 0 \text{ mod } I.$$

But we also have $\delta \mathbf{x}_j = Q_j \text{ mod } I$, and consequently

$$\delta^s P_j(\mathbf{x}', \mathbf{x}_j) = \delta^s P_j\left(\mathbf{x}', \frac{Q_j(\mathbf{x}', \mathbf{x}_{d+1})}{\delta}\right) \text{ mod } I$$

for $d+2 \leq j \leq n$ and s large enough. If we now increase q in the formula (1) above we obtain

$$\delta^q f = R(\delta \mathbf{x}_{d+2}, \ldots, \delta \mathbf{x}_n) \text{ mod } I,$$

where $R \in \mathcal{O}_{d+1}[\mathbf{y}_{d+2}, \ldots, \mathbf{y}_n]$. Rewriting this latter polynomial in powers of

$$\mathbf{y}_{d+2} = \delta \mathbf{x}_{d+2} - Q_{d+2}, \ldots, \mathbf{y}_n = \delta \mathbf{x}_n - Q_n,$$

since $\delta \mathbf{x}_j - Q_j \in I$ for $d+2 \leq j \leq n$, we get

$$\begin{aligned}\delta^q f &= R(\delta \mathbf{x}_{d+2}, \ldots, \delta \mathbf{x}_n) = \\ &= R(Q_{d+2}, \ldots, Q_n) + \sum_{|\nu| \geq 1} c_\nu (\delta \mathbf{x}_{d+2} - Q_{d+2})^{\nu_{d+2}} \cdots (\delta \mathbf{x}_n - Q_n)^{\nu_n} = \\ &= R(Q_{d+2}, \ldots, Q_n) \text{ mod } I.\end{aligned}$$

Finally, dividing by $P_{d+1} \in I$ we obtain

$$\delta^q f = R^* \text{ mod } I,$$

where $R^* \in \mathcal{O}_d[\mathbf{x}_{d+1}]$ is a polynomial of degree $< p$. Since by hypothesis $f \in \mathfrak{p}$ we deduce that $R^* \in \mathfrak{p}$ or, taking classes mod $\mathfrak{p}$:

$$R^*(\mathbf{x}', \theta_{d+1}) = 0.$$

This is only possible if R^* is identically zero, since its degree is strictly smaller than the degree p of the irreducible polynomial P_{d+1} of θ_{d+1}. Whence

$$\delta^q f = 0 \text{ mod } I,$$

which is exactly what we wanted. □

As a consequence of Rückert's Parametrization Theorem we obtain:

Proposition 3.5 *(Abhyankar's Parametrization) Let $\mathfrak{p}$ be a prime ideal of $\mathcal{O}_n$ (resp. $\mathcal{F}_n$). Then the localization $(\mathcal{O}_n)_\mathfrak{p}$ (resp. $(\mathcal{F}_n)_\mathfrak{p}$) is a regular local ring.*

Proof. Indeed, using the notations of Proposition 3.4, the ideal I generates $\mathfrak{p}(\mathcal{O}_n)_\mathfrak{p}$, since $\delta \notin \mathfrak{p}$, and consequently it is a unit in $(\mathcal{O}_n)_\mathfrak{p}$. This means that

$$P_{d+1}, \delta x_{d+2} - Q_{d+2}, \ldots, \delta x_n - Q_n$$

generate the maximal ideal of $(\mathcal{O}_n)_\mathfrak{p}$. Since

$$r = n - d = \operatorname{ht}(\mathfrak{p}) = \dim((\mathcal{O}_n)_\mathfrak{p}),$$

this latter localization is regular, and the elements above form a regular system of parameters. □

We finally give an alternative formulation of Rückert's Parametrization:

Proposition 3.6 *In the situation of Proposition 3.4 and with its notations, there is another ideal I' such that*

a) $I = I' \cap \mathfrak{p}$,

b) $\delta \in \sqrt{I'} \setminus \mathfrak{p}$, *and*

c) $\delta \mathfrak{p} \subset \sqrt{I}$.

Proof. Consider the primary decomposition of I ([A-McD §§4,7])

$$I = \mathfrak{q}_1 \cap \cdots \cap \mathfrak{q}_s.$$

Then we put

$$I' = \bigcap_{\delta \in \sqrt{\mathfrak{q}_i}} \mathfrak{q}_i, \qquad I'' = \bigcap_{\delta \notin \sqrt{\mathfrak{q}_i}} \mathfrak{q}_i.$$

Now, since $\delta^q \mathfrak{p} \subset I$, we have $\delta^q \mathfrak{p} \subset \mathfrak{q}_i$. But the $\mathfrak{q}_i$'s are primary ideals, and consequently either $\delta \in \sqrt{\mathfrak{q}_i}$ or $\mathfrak{p} \subset \mathfrak{q}_i$. Whence, $\mathfrak{p} \subset I''$. Thus $I \supset I' \cap \mathfrak{p}$, and the other inclusion is obvious. This shows a) and b), and c) follows at once:

$$\delta \mathfrak{p} \subset \sqrt{I'} \cap \mathfrak{p} = \sqrt{I}.$$

□

Remark 3.7 Under the same setting and notation as in Proposition 3.4, we have a canonical isomorphism

$$\mathcal{O}_d[x_{d+1}]/P_{d+1} \to \mathcal{O}_{d+1}/P_{d+1} = H,$$

and the inclusion $H \subset B$ induces another isomorphism

$$H_\delta \to B_\delta,$$

where the index δ means "ring of fractions whose denominators are powers of δ".

Indeed, both mappings are clearly injective. Moreover, the first one is onto by division by P_{d+1} in $\mathcal{O}_{d+1}$. Finally, the second mapping is also onto by Proposition 3.2. □

This fact is often quoted by saying that *every analytic domain is birationally equivalent to a hypersurface* (note that H is defined by a single equation).

4 Nagata's Jacobian Criteria

In this section we will denote $A = \mathcal{O}_n$ (resp. $\mathcal{F}_n$).

(4.1) Jacobians and regularity ideals. For any power series $f_1, \ldots, f_s \in A$ and indices $1 \leq i_1 < \cdots < i_s \leq n$ we have the *Jacobian of order s*

$$\frac{D(f_1, \ldots, f_s)}{D(\mathbf{x}_{i_1}, \ldots, \mathbf{x}_{i_s})} = \det\left(\frac{\partial f_j}{\partial \mathbf{x}_{i_\ell}}\right)_{1 \leq j, \ell \leq s}.$$

If I is an ideal of A, we will denote by $J_s(I)$ the ideal generated by I and all the above Jacobians of order s for $f_1, \ldots, f_s \in I$. If $h_1, \ldots, h_t$ generate I, then the elements

$$h_1, \ldots, h_t; \ \frac{D(h_{j_1}, \ldots, h_{j_s})}{D(\mathbf{x}_{i_1}, \ldots, \mathbf{x}_{i_s})}, \ 1 \leq j_1 < \cdots < j_s \leq t, \ 1 \leq i_1 < \cdots < i_s \leq n,$$

clearly generate $J_s(I)$. This ideal is called the *Jacobian ideal of order s of I*. It is easy to check that $J_s(I)$ is invariant by linear changes of coordinates.

On the other hand, we will denote by $G_s(I)$ the ideal generated by all the series $h \in A$ such that hI is contained in some ideal generated by s elements of I.

Finally, the ideal

$$R_s(I) = \sqrt{G_s(I) \cap J_s(I)}$$

is called the *regularity ideal of order s of I*.

Lemma 4.2 *Let $\mathfrak{p}$ a prime ideal of height $r > 0$ of A. There are then $h_1, \ldots, h_r \in \mathfrak{p}$ and a Jacobian*

$$\frac{D(h_1, \ldots, h_r)}{D(\mathbf{x}_{i_1}, \ldots, \mathbf{x}_{i_r})} \notin \mathfrak{p}.$$

Proof. After a linear change of coordinates we have $P_{d+1}, \ldots, P_n$ verifying the conditions of Proposition 3.3. From condition *c)* we get

$$\frac{D(P_{n-r+1}, \ldots, P_n)}{D(\mathbf{x}_{n-r+1}, \ldots, \mathbf{x}_n)} = \frac{\partial P_{n-r+1}}{\partial \mathbf{x}_{n-r+1}} \cdots \frac{\partial P_n}{\partial \mathbf{x}_n} \notin \mathfrak{p}.$$

Indeed, since P_j is the irreducible polynomial of $\mathbf{x}_j \bmod \mathfrak{p}$, and the degree of $\partial P_j/\partial \mathbf{x}_j$ is stricty smaller, we deduce that

$$\frac{\partial P_j}{\partial \mathbf{x}_j}(\mathbf{x}_j \bmod \mathfrak{p}) \neq 0,$$

that is, $\partial P_j/\partial \mathbf{x}_j \notin \mathfrak{p}$. □

The following criterion is of utmost importance:

Proposition 4.3 *(Regularity) Let* $\mathfrak{p}$ *a prime ideal of* A *and* $I \subset \mathfrak{p}$ *another ideal. The following assertions are equivalent:*

a) *The local ring* $A_\mathfrak{p}/IA_\mathfrak{p}$ *is regular of dimension* $\text{ht}(\mathfrak{p}) - s$.

b) $\mathfrak{p} \not\supset R_s(I)$.

c) $\mathfrak{p} \not\supset J_s(I)$ *and* $\text{ht}(IA_\mathfrak{p}) \leq s$.

Proof. We introduce some further notation:

$$B = A_\mathfrak{p}, \quad \mathfrak{n} = \mathfrak{p}A_\mathfrak{p}, \quad \mathfrak{a} = IA_\mathfrak{p},$$
$$\kappa = B/\mathfrak{n} = \text{quotient field of } A/\mathfrak{p},$$
$$r = \text{ht}(\mathfrak{p}) = \dim(B), \quad t = r - s.$$

$a) \Rightarrow b)$ First we will find elements $f_1, \ldots, f_s \in I$ and $f_{s+1}, \ldots, f_r \in \mathfrak{p}$ such that

$$\{f_1, \ldots, f_s\}B = \mathfrak{a}, \quad \{f_1, \ldots, f_r\}B = \mathfrak{n}.$$

In fact, since B is regular of dimension r (Proposition 3.4) and $B/\mathfrak{a}$ is regular of dimension $r - s$ (hypothesis a)), we have

$$r = \dim_\kappa(\mathfrak{n}/\mathfrak{n}^2) = \dim_\kappa(\mathfrak{n}/(\mathfrak{n}^2 + \mathfrak{a})) + \dim_\kappa((\mathfrak{n}^2 + \mathfrak{a})/\mathfrak{n}^2),$$
$$r - s = \dim_\kappa(\mathfrak{n}/(\mathfrak{n}^2 + \mathfrak{a}))$$

([A-McD 11.22]). Consequently, there are $f_1, \ldots, f_s \in I$ and $f_{s+1}, \ldots, f_r \in \mathfrak{p}$ such that $f_1, \ldots, f_r$ generate $\mathfrak{n}$ mod $\mathfrak{n}^2$. By Nakayama's Lemma, $\{f_1, \ldots, f_r\}B = \mathfrak{n}$, and we set $\mathfrak{b} = \{f_1, \ldots, f_s\}B \subset \mathfrak{a}$. We have a surjective homomorphism $B/\mathfrak{b} \to B/\mathfrak{a}$, and

$$r - s = \dim(B/\mathfrak{a}) \leq \dim(B/\mathfrak{b}) \leq r - s,$$

the last inequality because $f_{s+1}, \ldots, f_s$ generate the maximal ideal $\mathfrak{n}$ mod $\mathfrak{b}$. We conclude that $B/\mathfrak{b}$ is regular of dimension $r - s$. In particular $\mathfrak{b}$ is prime ([A-McD 11.23]). Now, if $\mathfrak{b} \neq \mathfrak{a}$, the dimension of $B/\mathfrak{b}$ would be $> r - s$, which is impossible. Thus

$$\{f_1, \ldots, f_s\}B = \mathfrak{b} = \mathfrak{a}$$

that is,

$$\{f_1, \ldots, f_s\}A_\mathfrak{p} = IA_\mathfrak{p}.$$

It follows that there is some $g \notin \mathfrak{p}$ with

$$gI \subset \{f_1, \ldots, f_s\}A \subset I,$$

and consequently

4 Nagata's Jacobian Criteria

$$G_s(I) \not\subset \mathfrak{p}.$$

On the other hand, let $h \in \mathfrak{p}$. There are $u \in A \setminus \mathfrak{p}$ and $g_1, \ldots, g_r \in A$ such that

$$uh = \sum_{j=1}^{r} g_j f_j.$$

Derivating this equality we get

$$u \frac{\partial h}{\partial x_i} + \frac{\partial u}{\partial x_i} h = \sum_{j=1}^{r} g_j \frac{\partial f_j}{\partial x_i} \quad \text{mod } f_1, \ldots, f_r,$$

and multiplying by u

$$u^2 \frac{\partial h}{\partial x_i} = \sum_{j=1}^{r} u g_j \frac{\partial f_j}{\partial x_i} \quad \text{mod } f_1, \ldots, f_r,$$

for $1 \leq i \leq n$ (note that $uh = 0 \bmod f_1, \ldots, f_r$).

On the other hand, by Lemma 4.2 there are $h_1, \ldots, h_r \in \mathfrak{p}$ and some Jacobian, say

$$\delta = \frac{D(h_1, \ldots, h_r)}{D(\mathbf{x}_1, \ldots, \mathbf{x}_r)} \notin \mathfrak{p},$$

which means that δ is a unit in $A_\mathfrak{p}$. But the previous formulas for $h = h_k$, $1 \leq k \leq r$, show that δ belongs to the ideal generated in $A_\mathfrak{p}$ by $f_1, \ldots, f_r$ and their Jacobians of order r. Hence some of these latter must be a unit in $A_\mathfrak{p}$, or in other words, must not belong to $\mathfrak{p}$. Computing that particular Jacobian through the rows corresponding to $f_1, \ldots, f_s$ we will find a new Jacobian of order s

$$\frac{D(f_1, \ldots, f_s)}{D(\mathbf{x}_{i_1}, \ldots, \mathbf{x}_{i_s})} \notin \mathfrak{p},$$

and consequently, $\mathfrak{p} \not\supset J_s(I)$.

Finally, since $\mathfrak{p}$ is a prime ideal, and it contains neither $G_s(I)$ nor $J_s(I)$, it cannot contain $R_s(I) = \sqrt{G_s(I) \cap J_s(I)}$. This is condition b), and the first implication of the statement is proved.

b) $\Rightarrow$ c) Since $\sqrt{J_s(I)} \supset R_s(I)$ and $\mathfrak{p}$ is prime, the first half of c) clearly follows from b).

Let us now estimate $\mathrm{ht}(\mathfrak{a})$. The same argument as above shows that $\mathfrak{p} \not\supset G_s(I)$, and thus there are $h \notin \mathfrak{p}$ and $g_1, \ldots, g_s \in I$ such that

$$hI \subset \{g_1, \ldots, g_s\} A.$$

As h is a unit of $A_\mathfrak{p}$ we deduce

$$\mathfrak{a} = \{g_1, \ldots, g_s\} B,$$

and consequently $\mathrm{ht}(\mathfrak{a}) \leq s$ ([A-McD 11.16]).

$c) \Rightarrow a)$ Since $J_s(I) \not\subset \mathfrak{p}$, there are $f_1, \ldots, f_s \in I$ and some Jacobian, say

$$\delta = \frac{D(f_1, \ldots, f_s)}{D(\mathbf{x}_1, \ldots, \mathbf{x}_s)},$$

which is a unit in B. We are to see that $f_1, \ldots, f_s \in \mathfrak{n}$ are linearly independent mod $\mathfrak{n}^2$, that is, their classes in the κ-linear space $\mathfrak{n}/\mathfrak{n}^2$ are linearly independent.

Indeed, suppose

$$a_1 f_1 + \cdots + a_s f_s = 0 \bmod \mathfrak{n}^2,$$

for some $a_1, \ldots, a_s \in B$. Derivating with respect to the $\mathbf{x}_i$'s we get

$$a_1 \frac{\partial f_1}{\partial \mathbf{x}_i} + \cdots + a_s \frac{\partial f_s}{\partial \mathbf{x}_i} = 0 \bmod \mathfrak{n}, \quad 1 \leq i \leq s,$$

(note that by the Leibnitz Formula any derivative of an element of $\mathfrak{n}^2$ belongs to $\mathfrak{n}$, and that the f_i's are in $I \subset \mathfrak{p} \subset \mathfrak{n}$). We thus have a system of linear equations in the unknowns $a_j, 1 \leq j \leq s$, with determinant $\delta \neq 0 \bmod \mathfrak{n}$. Hence this system has only the trivial solution over the field $\kappa = B/\mathfrak{n}$, that is, the a_j's are 0 mod $\mathfrak{n}$ as wanted.

Now, since B is a local regular ring (Proposition 3.5), we have

$$r = \dim(B) = \dim_\kappa(\mathfrak{n}/\mathfrak{n}^2)$$

([A-McD 11.22]). Therefore, we can extend $f_1, \ldots, f_s$ to a base $f_1, \ldots, f_r$ of $\mathfrak{n}/\mathfrak{n}^2$, with certain $f_{s+1}, \ldots, f_r \in \mathfrak{n}$. In other words, $\{f_1, \ldots, f_r\}B + \mathfrak{n}^2 = \mathfrak{n}$, and by Nakayama's Lemma

$$\{f_1, \ldots, f_r\}B = \mathfrak{n}.$$

We now set

$$\mathfrak{b}_\ell = \{f_1, \ldots, f_\ell\}B, \quad 1 \leq \ell \leq r,$$

and we have

$$\dim(B/\mathfrak{b}_\ell) = t \geq r - \ell.$$

Indeed, if $g_1, \ldots, g_t \in B$ give a regular system of parameters in the class ring $B/\mathfrak{b}_\ell$, then $g_1, \ldots, g_t, f_1, \ldots, f_\ell$ are a regular system of parameters of B; consequently, $r = \dim(B) \leq t + \ell$. As $f_{\ell+1}, \ldots, f_r$ generate $\mathfrak{n}$ mod $\mathfrak{b}_\ell$, we conclude that $B/\mathfrak{b}_\ell$ is local regular of dimension $r - \ell$. In particular we have the chain of prime ideals

$$(0) \subset \mathfrak{b}_1 \subset \cdots \subset \mathfrak{b}_s,$$

and so $\mathrm{ht}(\mathfrak{b}_s) \geq s$. But $\mathfrak{b}_s \subset \mathfrak{a}$ and $\mathrm{ht}(\mathfrak{a}) \leq s$, so that it must be $\mathfrak{b}_s = \mathfrak{a}$, and the local ring

$$A_\mathfrak{p}/IA_\mathfrak{p} = B/\mathfrak{b}_s$$

is regular of dimension $r - s = \mathrm{ht}(\mathfrak{p}) - s$. □

Next we apply the preceding criterion to the situation of Rückert's Parametrization Theorem to obtain the final formulation of parametrization in our setting:

4 Nagata's Jacobian Criteria

Proposition 4.4 *(Local Parametrization Theorem) Let $\mathfrak{p}$ be a prime ideal of height r of $\mathcal{O}_n$. After a linear change of coordinates the following conditions hold true:*

a) *The canonical homomorphism $\mathcal{O}_{n-r} \to B = \mathcal{O}_n/\mathfrak{p}$ is injective and finite.*

b) *The class $\theta = x_{n-r+1} \mod \mathfrak{p}$ is a primitive element of the quotient field L of $\mathcal{O}_n/\mathfrak{p}$ over the one K of $\mathcal{O}_{n-r}$.*

c) *The irreducible polynomial over K of the primitive element θ is a distinguished polynomial $P \in \mathcal{O}_{n-r}[x_{n-r+1}]$, whose discriminant is denoted by δ.*

d) *The canonical homomorphism $(\mathcal{O}_{n-r+1}/P)_\delta \to B_\delta$ is an isomorphism.*

e) *The localization B_δ is a regular ring: if $\mathfrak{q}$ is a prime ideal of B and $\delta \notin \mathfrak{q}$, the local ring $B_\mathfrak{q}$ is regular.*

Proof. The first four conditions in the statement have been already proved (Proposition 3.4 and Remark 3.7). The last one will follow from the Regularity Jacobian Criterion once we see that $\delta \in R_s(\mathfrak{p})$.

First, by Proposition 3.4 e) there is q such that

$$\delta^q \mathfrak{p} \subset \{P, \delta x_{n-r+2} - Q_{n-r+2}, \ldots, \delta x_n - Q_n\}\mathcal{O}_n \subset \mathfrak{p},$$

and this means that $\delta^q \in G_r(\mathfrak{p})$. On the other hand

$$\frac{D(P, \delta x_{n-r+2} - Q_{n-r+2}, \ldots, \delta x_n - Q_n)}{D(x_{n-r+1}, x_{n-r+2}, \ldots, x_n)} = \frac{\partial P}{\partial x_{n-r+1}} \delta^{r-1} \in J_r(\mathfrak{p}).$$

Now, since δ is the resultant of P and $\partial P/\partial x_{n-r+1}$, there are $f, g \in \mathcal{O}_{n-r}[x_{n-r+1}]$ such that

$$\delta = fP + g\frac{\partial P}{\partial x_{n-r+1}}$$

([L V.10]). Thus

$$\delta^r = \delta^{r-1}\left(fP + g\frac{\partial P}{\partial x_{n-r+1}}\right) \in \left\{P, \frac{\partial P}{\partial x_{n-r+1}}\delta^{r-1}\right\}\mathcal{O}_n \subset J_r(\mathfrak{p}).$$

Hence

$$\delta^{q+r} \in G_r(\mathfrak{p}) \cap J_r(\mathfrak{p})$$

and so

$$\delta \in \sqrt{G_r(\mathfrak{p}) \cap J_r(\mathfrak{p})} = R_r(\mathfrak{p}).$$

$\square$

Applying the preceding result to the maximal ideal we obtain:

Corollary 4.5 *The series $f_1, \ldots, f_n \in \mathfrak{m}_A$ generate the maximal ideal $\mathfrak{m}_A$ if and only if*
$$\frac{D(f_1, \ldots, f_n)}{D(\mathbf{x}_1, \ldots, \mathbf{x}_n)}(0) \neq 0.$$

Proof. In fact, set $I = \{f_1, \ldots, f_n\}A$. Then $I = \mathfrak{m}_A$ if and only if $A_{\mathfrak{p}}/IA_{\mathfrak{p}}$ is a field, if and only if $A_{\mathfrak{p}}/IA_{\mathfrak{p}}$ is regular of dimension 0. By Proposition 4.3 with $\mathfrak{p} = \mathfrak{m}_A$ and $s = \operatorname{ht}(\mathfrak{m}_A) = n$, the latter assertion is equivalent to $\mathfrak{m}_A \not\supset J_n(I)$ and $\operatorname{ht}(IA_{\mathfrak{p}}) \leq n$. As this inequality always holds we are done. □

The preceding corollary is nothing but the well-known Inverse Function Theorem. Of course it can be formulated in the standard way:

Proposition 4.6 *Set $\mathbf{x} = (\mathbf{x}_1, \ldots, \mathbf{x}_n)$, $\mathbf{y} = (\mathbf{y}_1, \ldots, \mathbf{y}_p)$.*

a) *(Inverse Function Theorem) Let $f = (f_1, \ldots, f_n) \in \mathbb{K}\{\mathbf{x}\}^n$ (resp. $\mathbb{K}[[\mathbf{x}]]^n$), with $f(0) = 0$. Suppose that*
$$\frac{D(f_1, \ldots, f_n)}{D(\mathbf{x}_1, \ldots, \mathbf{x}_n)}(0) \neq 0.$$
Then there is $g = (g_1, \ldots, g_n)$ with $g(0) = 0$ such that
$$f(g_1(\mathbf{x}), \ldots, g_n(\mathbf{x})) = \mathbf{x}.$$
Furthermore such a g is unique.

b) *(Implicit Functions Theorem) Let $f = (f_1, \ldots, f_p) \in \mathbb{K}\{\mathbf{x}, \mathbf{y}\}^p$ (resp. $\mathbb{K}[[\mathbf{x}, \mathbf{y}]]^p$), with $f(0, 0) = 0$. Suppose that*
$$\frac{D(f_1, \ldots, f_p)}{D(\mathbf{y}_1, \ldots, \mathbf{y}_p)}(0, 0) \neq 0.$$
Then there is $g = (g_1, \ldots, g_p)$ with $g(0) = 0$ such that
$$f(\mathbf{x}, g_1(\mathbf{x}), \ldots, g_p(\mathbf{x})) = 0.$$
Furthermore such a g is unique.

Proof. a) The homomorphism
$$\varphi : \mathbb{K}\{\mathbf{x}\} \to \mathbb{K}\{\mathbf{x}\} \ : \ \mathbf{x}_i \mapsto f_i$$
is onto by Corollaries 4.5 and 1.8. Counting dimensions, we see that the kernel of φ is a prime ideal of height 0, that is, the kernel is (0). Hence φ is an isomorphism, and the g_i's we are looking for are
$$g_i = \varphi^{-1}(\mathbf{x}_i), \quad 1 \leq i \leq n.$$

b) Set

4 Nagata's Jacobian Criteria

$$A = \mathbb{K}\{\mathbf{x}\}, \quad A' = \mathbb{K}\{\mathbf{x}, \mathbf{y}\}, \quad I = \{f_1, \ldots, f_p\}A'$$

and consider the homomorphism

$$\varphi : A \to A'/I \; : \; \mathbf{x}_i \mapsto \mathbf{x}_i \bmod I.$$

Since

$$\frac{D(\mathbf{x}_1, \ldots, \mathbf{x}_n, f_1, \ldots, f_p)}{D(\mathbf{x}_1, \ldots, \mathbf{x}_n, \mathbf{y}_1, \ldots, \mathbf{y}_p)}(0,0) \neq 0$$

the homomorphism φ is an isomorphism (same argument as in a)) and the solution now is

$$g_i = \varphi^{-1}(\mathbf{y}_i \bmod I), \quad 1 \leq i \leq p.$$

$\square$

(4.7) Application: roots of a unit. Consider a unit $u \in \mathbb{C}\{\mathbf{x}\}$ where $\mathbf{x} = (\mathbf{x}_1, \ldots, \mathbf{x}_n)$, that is, $u(0) = a \neq 0$. Fix an integer $p \geq 0$. For any p-th root c of a we can apply the Implicit Functions Theorem to the equation $(\mathbf{y} + c)^p = u$ and get $v = y(\mathbf{x}) + c \in \mathbb{C}\{\mathbf{x}\}$ such that

$$v^p = u, \quad v(0) = c.$$

A similar result holds in the real case, whenever c exists, that is, when $a > 0$ or p is odd.

We also deduce:

Let $f \in \mathbb{C}\{\mathbf{t}\}$ (resp. $\mathbb{R}\{\mathbf{t}\}$) a power series of order p. Then there is an isomorphism of $\mathbb{C}\{\mathbf{t}\}$ (resp $\mathbb{R}\{\mathbf{t}\}$) that transforms f in $\mathbf{t}^p$ (resp. $\pm \mathbf{t}^p$).

Indeed, consider the complex case. By the preceding remark we can write

$$f = \mathbf{t}^p u = (\mathbf{t}v)^p$$

with $v(0) \neq 0$. We then define an analytic homomorphism by

$$\varphi : \mathbf{t} \mapsto \mathbf{t}v.$$

This homomorphism is surjective because $\varphi(\mathbf{t})$ generates the maximal ideal and injective because it is finite. Hence φ is an isomorphism, and its inverse transforms f in $\mathbf{t}^p$, as wanted.

In the real case the argument is similar, writing $f = \pm \mathbf{t}^p u$ with $u(0) > 0$ so that there is $v \in \mathbb{R}\{\mathbf{t}\}$ with $f = \pm(\mathbf{t}v)^p$. $\square$

Of course all of this works the same for formal power series.

We end this section with another important Jacobian Criterion:

Proposition 4.8 *(Equidimensionality) Let $I \neq (0)$ be an ideal of A. The following assertions are equivalent:*

a) The ideal I is radical and every associated prime of I has height s.

b) There is an element $\delta \in R_s(I)$ which is not a zero divisor mod I.

Proof. $a) \Rightarrow b)$ Let
$$I = \mathfrak{p}_1 \cap \cdots \cap \mathfrak{p}_r, \quad \operatorname{ht}(\mathfrak{p}_i) = s, \ 1 \leq i \leq r.$$
We then have
$$IA_{\mathfrak{p}_i} = \mathfrak{p}_i A_{\mathfrak{p}_i}$$
and by the Regularity Jacobian Criterion (Proposition 4.3)
$$R_s(I) \not\subset \mathfrak{p}_i, \quad 1 \leq i \leq r.$$
Since the $\mathfrak{p}_i$'s are prime ideals, it follows
$$R_s(I) \not\subset \mathfrak{p}_1 \cup \cdots \cup \mathfrak{p}_r$$
([A-McD 1.11.i)]). Now the union of the $\mathfrak{p}_i$'s is exactly the set of the zero divisors mod I ([A-McD 4.7]). We have thus proved $b)$.

$b) \Rightarrow a)$ Let us see first that
$$\delta \in \sqrt{\operatorname{Ann}(\sqrt{I}/I)}.$$
Recall here that the *annihilator* $\operatorname{Ann}(M)$ of an A-module M is the ideal of A consisting of the elements $a \in A$ such that $am = 0$ for all $m \in M$. Now consider any prime ideal $\mathfrak{p}$ that contains $\operatorname{Ann}(\sqrt{I}/I)$; in particular $I \subset \mathfrak{p}$. Suppose then $\delta \notin \mathfrak{p}$. It would follow from Proposition 4.3 that the local ring $A_{\mathfrak{p}}/IA_{\mathfrak{p}}$ is regular and so $IA_{\mathfrak{p}}$ is a prime ideal ([A-McD 11.23]). Hence
$$IA_{\mathfrak{p}} = \sqrt{I}A_{\mathfrak{p}},$$
and there is some $u \notin \mathfrak{p}$ with $u\sqrt{I} \subset I$, that is, $u \in \operatorname{Ann}(\sqrt{I}/I) \setminus \mathfrak{p}$. This contradiction shows that δ belongs to all prime ideals containing $\operatorname{Ann}(\sqrt{I}/I)$, or in other words,
$$\delta \in \sqrt{\operatorname{Ann}(\sqrt{I}/I)}.$$
Consequently $\delta^p \sqrt{I} \subset I$ for some $p \geq 1$, and since δ is not a zero divisor mod I we conclude $\sqrt{I} \subset I$. Hence $\sqrt{I} = I$ and I is radical.

We can thus write $I = \mathfrak{p}_1 \cap \cdots \cap \mathfrak{p}_r$, where the $\mathfrak{p}_i$'s are prime. As δ is not a zero divisor, we already remarked that $\delta \notin \mathfrak{p}_1 \cup \cdots \cup \mathfrak{p}_r$. Then, and again by Proposition 4.3, the local ring $A_{\mathfrak{p}_i}/IA_{\mathfrak{p}_i}$ is regular of dimension $\operatorname{ht}(\mathfrak{p}_i) - s$. But since $IA_{\mathfrak{p}_i} = \mathfrak{p}_i A_{\mathfrak{p}_i}$, the ring $A_{\mathfrak{p}_i}/IA_{\mathfrak{p}_i}$ has dimension 0. We conclude that $0 = \operatorname{ht}(\mathfrak{p}_i) - s$, as wanted. □

5 Complexification

We consider in this section the extension of the ground field $- \otimes_\mathbb{R} \mathbb{C}$. In the case of analytic and formal rings this extension is specially well behaved and very useful to compare the complex and the real cases.

(5.1) Complexification. The tensor product $- \otimes_\mathbb{R} \mathbb{C}$ induces a covariant functor from the category of analytic (resp. formal) rings and homomorphisms over $\mathbb{R}$ onto that of analytic (resp. formal) rings and homomorphisms over $\mathbb{C}$. This functor will be called *complexification*, and we will use the notations

$$\widetilde{A} = A \otimes_\mathbb{R} \mathbb{C}, \quad \text{and} \quad \widetilde{\varphi} = \varphi \otimes_\mathbb{R} \mathbb{C}$$

for the complexification of a ring A and a homomorphism φ, respectively.

The explicit description of complexification is very natural. In order to present it we introduce a useful notion:

(5.2) Conjugation. The complexification of $A = \mathbb{R}\{\mathbf{x}\}$ (resp. $\mathbb{R}[[\mathbf{x}]]$) is $\widetilde{A} = \mathbb{C}\{\mathbf{x}\}$ (resp. $\mathbb{C}[[\mathbf{x}]]$), and it is canonically isomorphic to $A[\sqrt{-1}]$: every element $f = \sum a_\nu \mathbf{x}^\nu \in \widetilde{A}$ can be uniquely written in the form

$$f = \Re(f) + \sqrt{-1}\Im(f)$$

where

$$\Re(f) = \sum \Re(a_\nu)\mathbf{x}^\nu \in A, \quad \Im(f) = \sum \Im(a_\nu)\mathbf{x}^\nu \in A.$$

Consequently $A \subset \widetilde{A}$, and $\widetilde{A}$ is a rank 2 free A-module with basis $\{1, \sqrt{-1}\}$. In particular $\widetilde{A}$ is an integral extension of A.

Then we define an A-algebra homomorphism by

$$\widetilde{A} \to \widetilde{A} : f \mapsto \overline{f} = \Re(f) - \sqrt{-1}\Im(f).$$

This is an idempotent automorphism of $\widetilde{A}$ whose fixed part is exactly A, and which will be called quite obviously *conjugation*. (This is nothing but the automorphism deduced from the standard conjugation of $\mathbb{C}$ by applying the extension $A \otimes_\mathbb{R} -$.)

For every ideal $I \subset \widetilde{A}$ we have the *conjugated* ideal

$$\overline{I} = \{\overline{f} \mid f \in I\}.$$

One checks immediately that I is prime if and only if so is $\overline{I}$, and it follows that $\operatorname{ht}(I) = \operatorname{ht}(\overline{I})$ for arbitrary I.

(5.3) The extension $A \subset \widetilde{A}$ for $A = \mathbb{R}\{\mathbf{x}\}$ (resp. $\mathbb{R}[[\mathbf{x}]]$). For every ideal I of A we put $\widetilde{I} = I\widetilde{A}$. We collect in this paragraph the elementary properties of these *extended* ideals.

a) *Every extended ideal $\tilde{I}$ is invariant under conjugation. Conversely, for every ideal $J \subset \tilde{A}$ which is invariant under conjugation there is $I \subset A$ such that $J = \tilde{I}$: namely, $I = J \cap A$.*

Note that if J is invariant under conjugation and $h = \Re(h) + \sqrt{-1}\Im(h) \in J$, then also $\Re(h) - \sqrt{-1}\Im(h) \in J$. Consequently $\Re(h), \Im(h) \in J$. Hence J is generated by the $\Re(h)$'s and $\Im(h)$'s for $h \in J$, and those elements are in A. As a matter of fact, since $\Im(h) = -\Re(\sqrt{-1}h)$, we could even take only the $\Re(h)$'s as generators. □

b) *For every ideal $I \subset A$, we have $\tilde{I} \cap A = I$.*

If I is generated by $f_1, \ldots, f_s$ and $h \in \tilde{I}$, then there are $h_1, \ldots, h_s \in \tilde{A}$ such that
$$h = h_1 f_1 + \cdots + h_s f_s.$$
It follows
$$\Re(h) = \Re(h_1)f_1 + \cdots + \Re(h_s)f_s \in I,$$
and if $h \in A$, then $h = \Re(h) \in I$. □

c) *For any two ideals I, I' of A, we have $\widetilde{I \cap I'} = \tilde{I} \cap \tilde{I}'$. In particular, $I \subset I'$ if and only if $\tilde{I} \subset \tilde{I}'$.*

Arguing as above we see that if $h \in \tilde{I} \cap \tilde{I}'$ then $\Re(h) \in I \cap I'$, and similarly, $\Im(h) \in I \cap I'$. Thus
$$h \in \{\Re(h), \Im(h)\}\tilde{A} \subset \widetilde{I \cap I'}.$$
The other inclusion is evident. □

d) *For every ideal $I \subset A$, we have $\sqrt{\tilde{I}} = \sqrt{\tilde{I}}$. In particular, I is radical if and only if $\tilde{I}$ is radical.*

Indeed, $\sqrt{\tilde{I}}$ is invariant under conjugation, $\tilde{I}$ being so, and consequently we are reduced to see that $\sqrt{\tilde{I}} \cap A = \sqrt{I}$. But if $h \in A$ and $h^m \in \tilde{I}$, then $h^m \in \tilde{I} \cap A = I$. This gives one inclusion, and the other is obvious. □

e) *For every ideal $I \subset A$ and every $s \geq 1$, we have $\widetilde{J_s(I)} = J_s(\tilde{I})$.*

This follows immediately from the description of J_s by generators given in 4.1.

Concerning prime ideals the situation is a little more involved, but still well understood.

Proposition 5.4 *Set as above $A = \mathbb{R}\{x\}$ (resp. $\mathbb{R}[[x]]$), where $\mathbf{x} = (x_1, \ldots, x_n)$, and let $\mathfrak{p}$ be a prime ideal of A. There is then a prime ideal $\mathfrak{q}$ of $\tilde{A}$ that lies over $\mathfrak{p}$. This $\mathfrak{q}$ is unique up to conjugation, and*
$$\tilde{\mathfrak{p}} = \mathfrak{q} \cap \bar{\mathfrak{q}}.$$

Proof. In fact, by 5.3 *d)* the ideal $\widetilde{\mathfrak{p}}$ is radical, and consequently an intersection of prime ideals, say
$$\widetilde{\mathfrak{p}} = \mathfrak{q}_1 \cap \cdots \cap \mathfrak{q}_r.$$
Note now that this intersection must include the conjugates $\overline{\mathfrak{q}}_i$, because $\widetilde{\mathfrak{p}}$ is invariant under conjugation. Hence we can rewrite the equality above in the form
$$\widetilde{\mathfrak{p}} = (\mathfrak{q}_1 \cap \overline{\mathfrak{q}}_1) \cap \cdots \cap (\mathfrak{q}_s \cap \overline{\mathfrak{q}}_s).$$
Pick now for every $i = 1, \ldots, s$ an element
$$a_i \in \mathfrak{q}_i \setminus \bigcup_{j \neq i} \mathfrak{q}_j \cup \overline{\mathfrak{q}}_j$$
(after dropping the unnecessary primes). Then we get
$$(a_1 \overline{a}_1) \cdots (a_s \overline{a}_s) \in \widetilde{\mathfrak{p}} \cap A = \mathfrak{p}$$
and since $\mathfrak{p}$ is prime, some factor $a_i \overline{a}_i$, say $a_1 \overline{a}_1$, belongs to $\mathfrak{p}$. It follows that $a_1 \overline{a}_- \in \mathfrak{q}_j$ for all j, and consequently $a_1 \in \mathfrak{q}_j$ or $\overline{a}_1 \in \mathfrak{q}_j$. In any case we conclude $j = 1$, which proves the result. $\square$

Now we come back to the general situation. Suppose that we are given an analytic ring $A = \mathbb{R}\{\mathbf{x}\}/I$, where $\mathbf{x} = (\mathbf{x}_1, \ldots, \mathbf{x}_n)$; then $\widetilde{A} = \mathbb{C}\{\mathbf{x}\}/I\mathbb{C}\{\mathbf{x}\}$. From this it is clear that conjugation and extension of ideals as formulated in 5.2, 5.3 and Proposition 5.4 for $A = \mathbb{R}\{\mathbf{x}\}$ works the same for an arbitrary A. Of course, the analogous construction is valid in the formal case.

We thus deduce:

Proposition 5.5 *Let A be an analytic (resp. a formal) ring over $\mathbb{R}$. Then:*

a) $\dim(A) = \dim(\widetilde{A})$.

b) $\widetilde{A}$ *is reduced if and only if A reduced.*

c) $\widetilde{A}$ *is regular if and only if A regular.*

Proof. Part *a)* is equivalent to the assertion that $\operatorname{ht}(I) = \operatorname{ht}(\widetilde{I})$ for the ideal $I \subset \mathbb{R}\{\mathbf{x}\}$, and this follows from Proposition 5.4. Since a *reduced* ring is a ring such that $\sqrt{(0)} = (0)$, part *b)* is a consequence of 5.3 *d)*. Finally, part *c)* a consequence of Corollary 4.5 and part *a)*. $\square$

We now want to discuss when the complexification of a domain is also a domain. As should be clear from Proposition 5.4, some additional condition must be considered. To do that, we introduce a new notion:

Proposition and Definition 5.6 *Let A be a domain, K its quotient field and $\mathbf{t}$ an indeterminate. Consider the ring $A[\mathbf{t}]/(\mathbf{t}^2 + 1) = A[\iota]$, where ι stands for $\mathbf{t} \bmod \mathbf{t}^2 + 1$. The following assertions are equivalent:*

a) If $a^2 + b^2 = 0$ with $a, b \in A$, then $a = b = 0$.

b) The field K does not contain $\sqrt{-1}$.

c) The ring $A[\iota]$ is a domain.

In case these conditions hold the ring A is called 2-real.

Proof. The two first conditions are clearly equivalent. Suppose now *a)* holds, and let us prove that $A[\iota]$ is a domain.

Let $\alpha = a\mathbf{t} + b$, $\beta = c\mathbf{t} + d \in A[\mathbf{t}]$ be such that

$$\beta \neq 0 \bmod \mathbf{t}^2 + 1, \quad \alpha\beta = 0 \bmod \mathbf{t}^2 + 1.$$

The second congruence means that $\mathbf{t}^2 + 1$ divides the product $\alpha\beta$ in $A[\mathbf{t}]$. Since

$$\alpha\beta = (a\mathbf{t} + b)(c\mathbf{t} + d) = (ac)\mathbf{t}^2 + (ad + bc)\mathbf{t} + (bd),$$

we deduce $(ac)(\mathbf{t}^2 + 1) = \alpha\beta$, and consequently

$$ad + bc = 0, \quad ac - bd = 0.$$

This is a homogeneous system in the unknowns c, d with a non-trivial solution (it is $\beta \neq 0$). Thus

$$\det \begin{vmatrix} a & -b \\ b & a \end{vmatrix} = a^2 + b^2 = 0.$$

By *a)*, $a = b = 0$ and $\alpha = 0$. Thus $A[\iota]$ is a domain.

Conversely, suppose that $A[\iota]$ is a domain, and let $a, b \in A$ verify $a^2 + b^2 = 0$. We can then write $(a + \iota b)(a - \iota b) = 0$ in $A[\iota]$, and since $A[\iota]$ is a domain one of the factors is 0, say $a + \iota b = 0$. This means that $a + \mathbf{t}b \in (\mathbf{t}^2 + 1)A[\mathbf{t}]$, so that $a = b = 0$. We have hence proved *a)*. $\square$

The ring $A[\iota]$ gives another way of looking at complexifications:

Proposition 5.7 *Let A be an analytic (resp. a formal) ring over $\mathbb{R}$. There is then a canonical isomorphism $\widetilde{A} \simeq A[\iota]$.*

Proof. We define the homomorphism

$$\mathbb{R}\{\mathbf{x}\}[\mathbf{t}] \to \mathbb{C}\{\mathbf{x}\} : H(\mathbf{x}, \mathbf{t}) \mapsto H(\mathbf{x}, \sqrt{-1}),$$

and we need to check that

$$H(\mathbf{x}, \mathbf{t}) \in I[\mathbf{t}] + (\mathbf{t}^2 + 1) \quad \text{if and only if} \quad H(\mathbf{x}, \sqrt{-1}) \in \widetilde{I}$$

for any ideal $I \subset \mathbb{R}\{\mathbf{x}\}$. To that end, we first divide by $\mathbf{t}^2 + 1$ to get

$$H(\mathbf{x}, \mathbf{t}) = Q(\mathbf{x}, \mathbf{t})(\mathbf{t}^2 + 1) + a(\mathbf{x})\mathbf{t} + b(\mathbf{x}), \quad a, b \in \mathbb{R}\{\mathbf{x}\},$$

5 Complexification

and we see that we can suppose
$$H(\mathbf{x}, \mathbf{t}) = a(\mathbf{x})\mathbf{t} + b(\mathbf{x}), \ a, b \in \mathbb{R}\{\mathbf{x}\}.$$
In particular, by the properties of extended ideals (5.3), $H(\mathbf{x}, \sqrt{-1}) = a(\mathbf{x})\sqrt{-1} + b(\mathbf{x}) \in \widetilde{I}$ if and only if $a, b \in I$. We have thus to prove
$$a\mathbf{t} + b \in I[\mathbf{t}] + (\mathbf{t}^2 + 1) \ \text{ if and only if } \ a, b \in I$$
for any $a, b \in \mathbb{R}\{\mathbf{x}\}$. The "if" part is trivial, so let
$$a\mathbf{t} + b = (\alpha_0 \mathbf{t}^p + \cdots + \alpha_p) + (\beta_0 \mathbf{t}^q + \cdots + \beta_q)(\mathbf{t}^2 + 1)$$
with $\alpha_i \in I$. Equating coefficients we get $q = p - 2$ and

$$\begin{aligned}
\beta_0 &= -\alpha_0 \\
\beta_1 &= -\alpha_1 \\
\beta_2 &= -\alpha_2 - \beta_0 \\
&\vdots \\
\beta_{p-2} &= -\alpha_{p-2} - \beta_{p-4} \\
a &= \alpha_{p-1} + \beta_{p-3} \\
b &= \alpha_p + \beta_{p-2}
\end{aligned}$$

Looking at this system we succesively see that $\beta_1, \cdots, \beta_{p-2}$ belong to the ideal I, and finally that also $a, b \in I$, as wanted. $\square$

From the last two propositions we deduce:

Corollary 5.8 *Let A be an analytic (resp. a formal) ring over $\mathbb{R}$. Then the following assertions are equivalent:*

a) The complexification $\widetilde{A}$ is a domain.

b) A is a 2-real domain.

After this description of complexifications for rings we consider homomorphisms. Let $\varphi : A \to B$ be an analytic (resp. a formal) homomorphism. Suppose $A = \mathbb{R}\{\mathbf{x}\}/I$, $B = \mathbb{R}\{\mathbf{y}\}/J$, where $\mathbf{x} = (\mathbf{x}_1, \ldots, \mathbf{x}_n)$, $\mathbf{y} = (\mathbf{y}_1, \ldots, \mathbf{y}_p)$, and let φ be determined by
$$\varphi(\mathbf{x}_i \bmod I) = g_i(\mathbf{y}) \bmod J$$
for some $g_i \in \mathbb{R}\{\mathbf{y}\}$, $1 \leq i \leq n$ (Proposition 1.3). The complexification $\widetilde{\varphi}$ is determined then by
$$\widetilde{\varphi}(\mathbf{x}_i \bmod \widetilde{I}) = g_i(\mathbf{y}) \bmod \widetilde{J}$$
The same construction gives, of course, the complexification of a formal homomorphism.

Finally we leave to the reader the proof of the next result, which is an easy exercise involving the properties of extended ideals.

Proposition 5.9 Let $\varphi : A \to B$ be an analytic (resp. a formal) homomorphism. Then:

a) $\widetilde{\varphi}$ is injective (resp. surjective) if and only if φ is injective (resp. surjective).

b) $\widetilde{\varphi}$ is finite if and only if φ is finite.

III Normalization

Summary. This chapter is devoted to the study of integral closures in the categories of analytic and formal rings. The main goal is the description of normalizations. We then give two applications: uniformization of 1-dimensional rings, introducting quadratic transforms, and Newton-Puiseux's Theorem.

1 Integral Closures

We will denote by $\mathbb{K}' \supset \mathbb{K}$ either of the extensions: $\mathbb{C} = \mathbb{C}$, $\mathbb{C} \supset \mathbb{R}$, $\mathbb{R} = \mathbb{R}$. First of all we prove:

Lemma 1.1 *Let A be an analytic (resp. a formal) ring over $\mathbb{K}$ and $B \supset A$ an integral domain, which is a finite A-module. Then B is an analytic (resp. a formal) ring over $\mathbb{K}' \supset \mathbb{K}$.*

Proof. By Noether's Projection Lemma (Proposition II.2.6) we may assume $A = \mathbb{K}\{\mathbf{x}\}$, $\mathbf{x} = (\mathbf{x}_1, \ldots, \mathbf{x}_d)$; since B is a finite A-module, it is an integral extension of A. Let now $\mathfrak{n}$ be any maximal ideal of B. By the properties of integral extensions ([A-McD 5.8]) the ideal $\mathfrak{n} \cap A$ is the maximal ideal $\mathfrak{m}_A$ of A.

Consider an element $z \in B$. There is a monic polynomial $H \in A[\mathbf{z}]$ such that $H(z) = 0$. By the Preparation and Division Theorems we have $H = H' \cdot H''$, where H' is a distinguished polynomial and H'' a polynomial which is a unit in the ring of power series. This means

$$H'(\mathbf{x}, \mathbf{z}) = a_p + \cdots + a_1 \mathbf{z}^{p-1} + \mathbf{z}^p \in A[\mathbf{z}], \quad a_1, \ldots, a_p \in \mathfrak{m}_A$$

and

$$H''(\mathbf{x}, \mathbf{z}) = b_q + \cdots + b_1 \mathbf{z}^{q-1} + \mathbf{z}^q \in A[\mathbf{z}], \quad b_q \notin \mathfrak{m}_A.$$

Since B is a domain, there are two possibilities:

- $H'(z) = 0$. Then $z^p = -(a_p + \cdots + a_1 z^{p-1}) \in \mathfrak{m}_A B \subset \mathfrak{n}$.
- $H''(z) = 0$. Then $z(b_2 + \cdots + z^{q-1}) = -b_q$ is a unit in A, which implies that z is a unit in B.

This shows that B is a local ring, and $\mathfrak{n}$ is its maximal ideal. On the other hand the residue field $\mathbb{K}' = B/\mathfrak{n}$ is a finite extension of $A/\mathfrak{m} = \mathbb{K}$, and $\mathbb{K}' \supset \mathbb{K}$ is either $\mathbb{C} = \mathbb{C}$ or $\mathbb{C} \supset \mathbb{R}$ or $\mathbb{R} = \mathbb{R}$. We now

Claim. B *contains* $\mathbb{K}'\{\mathbf{x}\}$ *and is a finite* $\mathbb{K}'\{\mathbf{x}\}$*-module.*

Indeed, if $\mathbb{K}' = \mathbb{K}$, then $\mathbb{K}'\{\mathbf{x}\} = A$, and the assertion is trivial. We suppose, hence, $\mathbb{K}' = \mathbb{C}$ and $\mathbb{K} = \mathbb{R}$. We then consider the ring extension $B[\iota] \supset B$ described in Proposition and Definition II.5.6:

$$B[\iota] = B[\mathbf{t}]/(\mathbf{t}^2+1), \quad \iota = \mathbf{t} \bmod t^2+1,$$

where $\mathbf{t}$ is a new indeterminate.

The ring $B[\iota]$ is clearly a finite B-module and consequently a finite A-module, too. Hence, by the part already proved, $B[\iota]$ is a local ring whose maximal ideal $\mathfrak{n}'$ verifies $\mathfrak{n}' \cap B = \mathfrak{n}$. Hence the residue field of $B[\iota]$ is a finite extension of that of B, and so both are $\mathbb{C}$. Thus, there is some element $\iota' \in B$ whose residue class is $\sqrt{-1}$, which is also the residue class of ι. This can be expressed by $\iota - \iota' \in \mathfrak{n}'$, and we deduce

$$B[\iota] = B + \mathfrak{n}'.$$

On the other hand we have the following commutative triangle

$$\begin{array}{ccc} B/\mathfrak{n} = \mathbb{C} & \longrightarrow & \mathbb{C}[\mathbf{t}] \\ & \varphi \searrow \quad \swarrow \phi & \\ & B[\iota]/\mathfrak{n}B[\iota] & \end{array}$$

where φ is the homomorphism induced by the inclusion $B \subset B[\iota]$, and ϕ the extension of φ given by $\mathbf{t} \mapsto \iota$. Clearly ϕ is surjective, and we get an isomorphism

$$\mathbb{C}[\mathbf{t}]/I \simeq B[\iota]/\mathfrak{n}B[\iota],$$

where I is the kernel of ϕ. As $\mathbf{t}^2+1 \in I$ and $I \subset \mathbb{C}[\mathbf{t}]$ is a principal ideal, the generator of I must be one of the following three polynomials:

$$\mathbf{t}^2+1, \quad \mathbf{t}-\sqrt{-1}, \quad \mathbf{t}+\sqrt{-1}.$$

But if I were generated by the first one, we would have $\mathbb{C}[\mathbf{t}]/I = \mathbb{C} \oplus \mathbb{C}$, which is not a local ring, while $B[\iota]/\mathfrak{n}B[\iota]$ is. Thus $I = (\mathbf{t} \pm \sqrt{-1})$. In any case $\mathbb{C}[\mathbf{t}]/I = \mathbb{C}$ and the ring $B[\iota]/\mathfrak{n}B[\iota]$ is a field. Hence $\mathfrak{n}B[\iota]$ is a maximal ideal and must coincide with $\mathfrak{n}'$. We deduce

$$B[\iota] = B + \mathfrak{n}' = B + \mathfrak{n}B[\iota]$$

and by Nakayama's Lemma, $B = B[\iota]$. This means that $\iota \in B$, or in other words, that $\sqrt{-1} \in B$.

Finally,

$$\mathbb{K}'\{\mathbf{x}\} = \mathbb{C}\{\mathbf{x}\} = \mathbb{R}\{\mathbf{x}\}[\sqrt{-1}] = A[\sqrt{-1}] \subset B$$

and the claim is proved.

1 Integral Closures

We are now ready to show that B is an analytic ring over K'. First of all, by the claim there are finitely many elements $y_1, \ldots, y_s \in \mathfrak{n}$ such that

$$B = \mathbb{K}'\{\mathbf{x}\}[y_1, \ldots, y_s].$$

Furthermore, we know that every y_i is a root of some distinguished polynomial $P_i(\mathbf{x}, \mathbf{y}_i) \in \mathbb{K}'\{\mathbf{x}\}[\mathbf{y}_i]$. Using these y_i's we will define a surjective homomorphism of $\mathbb{K}'$-algebras

$$\pi : \mathbb{K}'\{\mathbf{x}, \mathbf{y}\} \to B, \quad \mathbf{y} = (\mathbf{y}_1, \ldots, \mathbf{y}_s).$$

To that end, after division by the P_i's, we can write any $f \in \mathbb{K}'\{\mathbf{x}, \mathbf{y}\}$ in the form:

$$f = \sum_{|\nu| \leq p} a_\nu \mathbf{y}_1^{\nu_1} \cdots \mathbf{y}_s^{\nu_s} + \lambda_1 P_1 + \cdots + \lambda_s P_s,$$

where p is any integer bigger than the sum of the degrees of the P_i's, and $\lambda_i \in \mathbb{K}'\{\mathbf{x}, \mathbf{y}\}$, $a_\nu \in \mathbb{K}'\{\mathbf{x}\}$. Quite obviously we set

$$\pi(f) = \sum_{|\nu| \leq p} a_\nu y_1^{\nu_1} \cdots y_s^{\nu_s} \in \mathbb{K}'\{\mathbf{x}\}[y_1, \ldots, y_s] = B.$$

With such a definition the conclusion would be immediate. Hence we need to show the validity of the above definition. That is, we need to check that for any element

$$f \in \mathbb{K}'\{\mathbf{x}\}[\mathbf{y}] \cap \{P_1, \ldots, P_s\} \mathbb{K}'\{\mathbf{x}, \mathbf{y}\},$$

it holds

$$f(\mathbf{x}, y_1, \ldots, y_s) = 0$$

(note that this substitution is always possible because f is a polynomial in the $\mathbf{y}_i$'s). But, if

$$f = \lambda_1 P_1 + \cdots + \lambda_s P_s, \quad \lambda_i \in \mathbb{K}'\{\mathbf{x}, \mathbf{y}\},$$

we can write

$$f = \lambda_1^{(r)} P_1 + \cdots + \lambda_s^{(r)} P_s + g^{(r)},$$

with

$$\lambda_i^{(r)} \in \mathbb{K}'\{\mathbf{x}\}[\mathbf{y}] \text{ and } g^{(r)} \in \{\mathbf{y}^\nu \,|\, |\nu| = r\} \mathbb{K}'\{\mathbf{x}, \mathbf{y}\}.$$

Thus

$$g^{(r)} = f - (\lambda_1^{(r)} P_1 + \cdots + \lambda_s^{(r)} P_s) \in \mathbb{K}'\{\mathbf{x}\}[\mathbf{y}],$$

and, since the Taylor expansion of $g^{(r)}$ as a series in $\mathbf{y}_1, \ldots, \mathbf{y}_s$ starts with monomials $\mathbf{y}^\nu$ with $|\nu| \geq r$, we get

$$g^{(r)} \in \{\mathbf{y}^\nu \,|\, |\nu| = r\} \mathbb{K}'\{\mathbf{x}\}[\mathbf{y}].$$

We now consider the ideals

$$I = \{P_1, \ldots, P_s\}\mathbb{K}'\{\mathbf{x}\}[\mathbf{y}] \subset \{\mathbf{x}_1, \ldots, \mathbf{x}_d, \mathbf{y}_1, \ldots, \mathbf{y}_s\}\mathbb{K}'\{\mathbf{x}\}[\mathbf{y}] = \mathfrak{p}$$

and by the preceding remarks we have

$$f \in \bigcap_{r \geq 0}(I + \mathfrak{p}^r).$$

Consequently, by Krull's Theorem applied to the ring

$$\mathbb{K}'\{\mathbf{x}\}[\mathbf{y}]_\mathfrak{p}/I\mathbb{K}'\{\mathbf{x}\}[\mathbf{y}]_\mathfrak{p},$$

there are elements $h \in \mathbb{K}'\{\mathbf{x}\}[\mathbf{y}] \setminus \mathfrak{p}$, $\alpha_1, \ldots, \alpha_s \in \mathbb{K}'\{\mathbf{x}\}[\mathbf{y}]$ such that

$$hf = \alpha_1 P_1 + \cdots + \alpha_s P_s.$$

As this expression only involves polynomials in the $\mathbf{y}_i$'s, the substitution $\mathbf{y}_i = y_i$ is allowed, and so we get

$$h(\mathbf{x}, y_1, \ldots, y_s)f(\mathbf{x}, y_1, \ldots, y_s) = 0.$$

Finally, $h(\mathbf{x}, y_1, \ldots, y_s) \neq 0$. Indeed, otherwise we would deduce

$$h(\mathbf{x}, 0) \in (\{y_1, \ldots, y_s\}B) \cap \mathbb{K}'\{\mathbf{x}\} \subset \mathfrak{p} \cap \mathbb{K}'\{\mathbf{x}\} = \{\mathbf{x}_1, \ldots, \mathbf{x}_d\}\mathbb{K}'\{\mathbf{x}\}$$

and so $h(0,0) = 0$, which is impossible, because

$$h(\mathbf{x}, \mathbf{y}) \notin \mathfrak{p} = \{\mathbf{x}_1, \ldots, \mathbf{x}_d, \mathbf{y}_1, \ldots, \mathbf{y}_s\}\mathbb{K}'\{\mathbf{x}\}[\mathbf{y}].$$

The proof is now complete. □

Remarks 1.2 The coefficient field $\mathbb{K}'$ can be determined in some cases, but not in general.

a) A domain which is a finite module over an analytic (resp. a formal) ring over $\mathbb{C}$ is again an analytic (resp. a formal) ring over $\mathbb{C}$.

Indeed, in this case $\mathbb{K} = \mathbb{C}$, and consequently $\mathbb{K}' = \mathbb{C}$, too. □

b) A 2-real domain which is a finite module over an analytic (resp. a formal) ring over $\mathbb{R}$ is again an analytic (resp. a formal) ring over $\mathbb{R}$.

Note that a 2-real domain cannot contain $\sqrt{-1}$, and so the coefficient field $\mathbb{K}'$ must be $\mathbb{R}$. □

c) Consider the analytic domain $A = \mathbb{R}\{\mathbf{x}_1, \mathbf{x}_2\}/(\mathbf{x}_1^2 + \mathbf{x}_2^2)$. It is not 2-real, but its coefficient field is $\mathbb{R}$.

d) With A as in *c)* denote by K the quotient field of A; put $x_1 = \mathbf{x}_1 \mod \mathbf{x}_1^2 + \mathbf{x}_2^2$, $x_2 = \mathbf{x}_2 \mod \mathbf{x}_1^2 + \mathbf{x}_2^2$. The element $x/y \in K$ is then integral over A, $(x/y)^2 = x^2/y^2 = -1$, and the domain $B = A[x/y]$ is a finite A-module and it is not 2-real. Thus B is an analytic ring, but now its coefficient field is $\mathbb{C}$ instead of $\mathbb{R}$.

From Proposition 1.1 we deduce the key result concerning integral closures:

Proposition 1.3 *Let A be an analytic (resp. a formal) ring over $\mathbb{K}$, and suppose that A is a domain with quotient field K. Let L be a finite field extension of K and B the integral closure of A in L. Then B is a finite A-module and an analytic (resp. a formal) ring over $\mathbb{K}' \supset \mathbb{K}$.*

Proof. Choose any primitive element θ of L over K. If
$$a_0 \theta^p + a_1 \theta^{p-1} + \cdots + a_p = 0$$
with $a_i \in A$, $a_0 \neq 0$, the element $a_0 \theta$ is also a primitive element and it is integral over A:
$$(a_0 \theta)^p + a_1 a_0 (a_0 \theta)^{p-1} + \cdots + a_p a_0^p = 0.$$
We thus may merely assume that θ is integral over A, so that $A[\theta] \subset B$. Furthermore, $A[\theta]$ is a finite A-module ([A-McD 5.1]) and by Proposition 1.1 $A[\theta]$ is an analytic ring. We note now that B is the integral closure of $A[\theta]$ in its quotient field $L = K[\theta]$, and by Corollary II.3.3, B is a finite $A[\theta]$-module. By Proposition 1.1 again, B is an analytic ring. □

2 Normalization

This section is devoted to the following notion:

Definition 2.1 *Let A be a reduced analytic (resp. formal) ring over $\mathbb{K}$ and K its total ring of fractions.*

a) *The integral closure of A in K is called the* normalization *of A and denoted by A^ν.*

b) *A is called* normal *if it is integrally closed in K.*

In order to understand this better we need a precise description of the normalization.

Proposition 2.2 *Let A be a reduced analytic (resp. formal) ring over $\mathbb{K}$. Consider the associated primes $\mathfrak{p}_1, \ldots, \mathfrak{p}_s$ of (0) and put $A_i = A/\mathfrak{p}_i$. There is then a canonical isomorphism*
$$A^\nu \simeq A_1^\nu \times \cdots \times A_s^\nu.$$
In particular, A^ν has finitely many maximal ideals $\mathfrak{m}_i^\nu$, and

a) *A_i^ν is the localization of A^ν at $\mathfrak{m}_i^\nu$.*

b) *$\mathfrak{p}_i$ is the kernel of the canonical homomorphism $A \to A^\nu_{\mathfrak{m}_i^\nu}$.*

Proof. The total ring of fractions K of A consists of all fractions whose denominators are not zero divisors, and so the canonical homomorphism $A \to K$ is an inclusion. Furthermore, the zero divisors of A are exactly the elements of the associated primes $\mathfrak{p}_i$ ([A-McD 4.7]). Thus K is the semilocalization $A_{\mathfrak{p}_1,\ldots,\mathfrak{p}_s}$ and $\mathfrak{p}_1 K, \ldots, \mathfrak{p}_s K$ are the maximal ideals of K. Setting $K_i = K/\mathfrak{p}_i K$ for $1 \le i \le s$, we obtain a canonical homomorphism

$$K \to K_1 \times \cdots \times K_s.$$

This homomorphism is actually an isomorphism, by the famous Chinese Remainder Theorem ([A-McD 1.10]). Indeed, we have

$$K = \mathfrak{p}_i K + \bigcap_{j \ne i} \mathfrak{p}_j K, \quad \text{and} \quad \bigcap_i \mathfrak{p}_i K = (0)$$

for all i's.

On the other hand,

$$K_i = K/\mathfrak{p}_i K = \text{quotient field of } A/\mathfrak{p}_i$$

and we have a commutative square

$$\begin{array}{ccc} A & \longrightarrow & K \\ \downarrow & & \downarrow \tau \\ A/\mathfrak{p}_1 \times \cdots \times A/\mathfrak{p}_s & \longrightarrow & K_1 \times \cdots \times K_s \end{array}$$

Here the two horizontal arrows are inclusions, and the right vertical one is an isomorphism, as remarked above.

After this preparation we prove the formula of the statement. It is clear from the construction that τ maps A^ν into $A_1^\nu \times \cdots \times A_s^\nu$ and we claim that this inclusion is actually the isomorphism

$$A^\nu \simeq A_1^\nu \times \cdots \times A_s^\nu$$

we sought. Indeed, by Proposition 1.3 every A_i^ν is a finite A_i-module, and consequently a finite A-module (the homomorphism $A \to A_i$ is onto). Hence $A_1^\nu \times \cdots \times A_s^\nu$ is a finite A-module, too. Let a be an element of that product. Since A is noetherian, the submodule

$$A[a] \subset A_1^\nu \times \cdots \times A_s^\nu$$

is also a finite A-module, and a is integral over A. This shows that τ is surjective, and the proof is complete. $\square$

On the way we have shown:

Proposition 2.3 *The normalization A^ν of a reduced analytic (resp. formal) ring A over $\mathbb{K}$ is a finite A-module.*

2 Normalization

We also get:

Proposition 2.4 *A normal analytic ring over $\mathbb{K}$ is an integral domain.*

Proof. With the notations of Proposition 2.4, if A is normal then
$$A = A^\nu = A_1^\nu \times \cdots \times A_s^\nu.$$
But since A is a local ring, this product cannot have more than one factor, and consequently (0) is a prime ideal, and A is a domain. □

We now want to control the behaviour of the coefficient field under normalization. As usual the complex case is simpler:

Proposition 2.5 *Let A be a reduced analytic (resp. formal) ring over $\mathbb{C}$. The normalization A^ν of A is a finite product of normal analytic rings over $\mathbb{C}$ and the correspondence*
$$\mathfrak{p} \mapsto (A/\mathfrak{p})^\nu$$
gives a bijection between the associated primes of $(0) \subset A$ and the factors of that product.

Proof. With the notations of Proposition 2.2 again, we have $A^\nu = A_1^\nu \times \cdots \times A_s^\nu$, and the A_i^ν's are analytic rings over $\mathbb{C}$ by Remark 1.2 a). □

The corresponding result in the real case is the following:

Proposition 2.6 *Let A be a reduced analytic (resp. formal) ring over $\mathbb{R}$. The normalization A^ν of A is a finite product of normal analytic rings over $\mathbb{R}$ and/or over $\mathbb{C}$, and the correspondence*
$$\mathfrak{p} \mapsto (A/\mathfrak{p})^\nu$$
gives a bijection between the associated primes of $(0) \subset A$ and the factors of that product.

Under this bijection the 2-real domains $A/\mathfrak{p}$ correspond to the analytic rings over $\mathbb{R}$.

Proof. Suppose first that the domain $A/\mathfrak{p}$ is 2-real. Since this notion only depends on the quotient field of the given domain (Proposition and Definition II.5.6) we see that $(A/\mathfrak{p})^\nu$ is also 2-real. Thus by Remark 1.2 b) the coefficient field of the latter is $\mathbb{R}$.

Let now $\mathfrak{p}$ an associated prime such that $A/\mathfrak{p}$ is not 2-real. Then the quotient field of that domain contains $\sqrt{-1}$ (Proposition and Definition II.5.6 again), and consequently $\sqrt{-1} \in (A/\mathfrak{p})^\nu$. Hence the coefficient field of that ring is $\mathbb{C}$. □

Finally concerning complexification we have:

Proposition 2.7 *Let A be a reduced analytic (resp. formal) ring over $\mathbb{R}$. Consider the associated primes $\mathfrak{p}_1,\ldots,\mathfrak{p}_r, \mathfrak{q}_1,\ldots,\mathfrak{q}_s$ of (0), so ordered that the domains $A_i/\mathfrak{p}_i$, $1 \le i \le r$, are 2-real, and the domains $B_j = A/\mathfrak{q}_j$, $1 \le j \le s$, are not. Then the canonical isomorphism*

$$A^\nu \simeq A_1^\nu \times \cdots \times A_r^\nu \times B_1^\nu \times \cdots \times B_s^\nu$$

extends to the complexification in the form

$$(\widetilde{A})^\nu \simeq \widetilde{A_1^\nu} \times \cdots \times \widetilde{A_r^\nu} \times (B_1^\nu \times B_1^\nu) \times \cdots \times (B_s^\nu \times B_s^\nu).$$

Proof. Let $\widetilde{A}$ be the complexification of A. By Propositions II.5.4 and II.5.8 the associated primes of $\widetilde{A}$ are

$$\mathfrak{p}_1\widetilde{A},\ldots,\mathfrak{p}_r\widetilde{A}$$

and

$$\mathfrak{n}_1,\overline{\mathfrak{n}}_1,\ldots,\mathfrak{n}_s,\overline{\mathfrak{n}}_s,$$

where every $\mathfrak{n}_j$ is a prime ideal of $\widetilde{A}$ such that $\mathfrak{n}_j \cap A = \mathfrak{q}_j$, and also $\overline{\mathfrak{n}}_j \cap A = \mathfrak{q}_j$. Hence, by Proposition 2.5 the ring $(\widetilde{A})^\nu$ is canonically isomorphic to the product

$$(\widetilde{A}/\mathfrak{p}_1\widetilde{A})^\nu \times \cdots \times (\widetilde{A}/\mathfrak{p}_r\widetilde{A})^\nu \times (\widetilde{A}/\mathfrak{n}_1)^\nu \times (\widetilde{A}/\overline{\mathfrak{n}}_1)^\nu \times \cdots \times (\widetilde{A}/\mathfrak{n}_s)^\nu \times (\widetilde{A}/\overline{\mathfrak{n}}_s)^\nu.$$

Thus, the assertion of the statement splits into the following:

a) $\widetilde{A_i^\nu} \simeq (\widetilde{A}/\mathfrak{p}_i\widetilde{A})^\nu$ for $i=1,\ldots,r$.

b) $B_j \simeq (\widetilde{A}/\mathfrak{n}_j)^\nu \simeq (\widetilde{A}/\overline{\mathfrak{n}}_j)^\nu$ for $j=1,\ldots,s$.

We start by proving a). First of all we note that $\widetilde{A}/\mathfrak{p}_i\widetilde{A}$ is the complexification of A_i. This domain is 2-real, and consequently its quotient field K_i does not contain $\sqrt{-1}$. Thus by Propositions II.5.6 and II.5.7 we have

$$\widetilde{A}/\mathfrak{p}_i\widetilde{A} = A_i[\sqrt{-1}], \quad \widetilde{A_i^\nu} = A_i^\nu[\sqrt{-1}],$$

and the quotient fields of the two rings $\widetilde{A}/\mathfrak{p}_i\widetilde{A}$ and $\widetilde{A_i^\nu}$ coincide both with the field $K_i[\sqrt{-1}]$. We have to see, hence, that $A_i^\nu[\sqrt{-1}]$ is the integral closure of $A_i[\sqrt{-1}]$ in $K_i[\sqrt{-1}]$. To do so, note that $A_i^\nu[\sqrt{-1}]$ is an integral extension of A_i^ν, and consequently of $A_i \subset A_i[\sqrt{-1}]$. Thus it remains to see that $A_i^\nu[\sqrt{-1}]$ is integrally closed in $K_i[\sqrt{-1}]$.

Let $f,g \in K_i$ be such that $h = f + \sqrt{-1}g$ is integral over $A_i^\nu[\sqrt{-1}]$. Since the latter domain is integral over A_i, the element h is a root of some monic polynomial $P \in A_i[\mathfrak{t}]$:

$$P(f + \sqrt{-1}) = 0.$$

As P is invariant by conjugation in $\widetilde{A_i} = A_i[\sqrt{-1}]$

$$P(f - \sqrt{-1}) = 0,$$

and consequently also $\overline{h} = f - \sqrt{-1}g$ is integral over A_i. Thus the elements

$$f = \frac{1}{2}(h + \overline{h}), \quad g = \frac{1}{2}(h - \overline{h})$$

are integral over A_i, and so they belong to A_i^ν. We conclude that $h \in A_i^\nu[\sqrt{-1}]$, as wanted.

We now show *b)*. Set

$$L_j = \text{quotient field of } B_j, \quad F_j = \text{quotient field of } \widetilde{A}/\mathfrak{n}_j.$$

Since $\mathfrak{n}_j \cap A = \mathfrak{q}_j$, we have an inclusion $B_j \subset \widetilde{A}/\mathfrak{n}_j$, which extends to another $L_j \subset F_j$. We claim that in fact $L_j = F_j$. Every element $h \in \widetilde{A}/\mathfrak{n}_j$ can be written in the form $h = f + \sqrt{-1}g$ with $f, g \in B_j$, and $\sqrt{-1} \in L_j$. We thus have $B_j \subset \widetilde{A}/\mathfrak{n}_j \subset F_j = L_j$, and since the first inclusion here is an integral extension, we get $B_j^\nu \simeq (\widetilde{A}/\mathfrak{n}_j)^\nu$ as wanted. As this argument also works for $\overline{\mathfrak{n}}_j$, the proof of *b)* is finished.

Whence, as remarked before, also the proof of the proposition is finished. □

We end the section with an easy corollary of the last result:

Corollary 2.8 *An analytic (resp. a formal) ring A over $\mathbb{R}$ is normal if and only if its complexification is normal.*

Proof. After the preceding proposition we only have to show that if $\widetilde{A}$ is normal, then there is no B_j. But if $\widetilde{A}$ is normal, then $\widetilde{A}$ is a domain (Proposition 2.4), and consequently A is a 2-real domain by Proposition II.5.8. □

3 Multiplicity in Dimension 1

In this section we consider some special properties of 1-dimensional rings. As usual, we only describe the analytic case, the formal one being analogous.

From now on, A stands for a reduced analytic ring of dimension 1 over $\mathbb{K} = \mathbb{R}$ or $\mathbb{C}$. We denote by K the total ring of fractions of A, and by $\mathfrak{m}$ its maximal ideal. Furthermore, since A is reduced, we have the decomposition $(0) = \mathfrak{p}_1 \cap \cdots \cap \mathfrak{p}_s$, where $\mathfrak{p}_1, \ldots, \mathfrak{p}_s$ are the minimal primes of A. We put $A_i = A/\mathfrak{p}_i$, $1 \leq i \leq s$ and call these A_i's the *branches* of A; we denote by K_i the quotient field of A_i, and by $\mathfrak{m}_i$ its maximal ideal. We also know that K is canonically isomorphic to $K_1 \times \cdots \times K_s$.

Finally we will denote by $\mathbb{K}(\{\mathfrak{t}\})$ the quotient field of $\mathbb{K}\{\mathfrak{t}\}$.

By Proposition II.2.3, any non-zero element $x \in \mathfrak{m}_i$ gives a unique non-trivial analytic homomorphism $\varphi : \mathbb{K}\{\mathbf{t}\} \to A_i$ such that $\mathbf{t} \mapsto x$ and conversely. Since A_i is a domain, such a homomorphism is injective, and by Example II.1.6 it is finite. We thus obtain a finite field extension $\mathbb{K}(\{\mathbf{t}\}) \subset K_i$ whose degree will be denoted by $\deg(x)$ or $\deg(\varphi)$.

Definition 3.1 *For every $i = 1, \ldots, s$, let $\mu_i(A)$ denote the smallest degree $\deg(\varphi)$ of a non-trivial analytic homomorphism $\varphi : \mathbb{K}\{\mathbf{t}\} \to A_i$. Then, the integer*

$$\mu(A) = \mu_1(A) + \cdots + \mu_s(A)$$

is called the multiplicity *of A.*

Our purpose now is to find a different computation for the multiplicity through the normalization B of A. We recall from Section 2 that B is the integral closure of A in K and that it is canonically isomorphic to the product $B_1 \times \cdots \times B_s$ where every B_i is the integral closure of A_i in K_i. We have:

Proposition 3.2 *Every B_i is an analytic ring over $\mathbb{K}'_i \supset \mathbb{K}$, isomorphic to $\mathbb{K}'_i\{\mathbf{t}\}$.*

Proof. By Propositions 2.6 and 2.7, B_i is an analytic ring over $\mathbb{K}'_i \supset \mathbb{K}_i$. Moreover, by Proposition 2.3, B_i is a finite A_i-module, and consequently the dimensions of B_i and A_i coincide. Thus B_i is a normal local ring of dimension 1, and so regular ([A-McD 9.2]). Whence B_i is isomorphic to $\mathbb{K}'_i\{\mathbf{t}\}$ by Lemma II.1.9. □

In what follows, we denote by $\mathfrak{n}_i$ the maximal ideal of B_i, and set

$$f_i = [\mathbb{K}'_i : \mathbb{K}].$$

Now, after fixing any isomorphism $\phi_i : B_i \simeq \mathbb{K}'_i\{\mathbf{t}\}$, we set

$$\omega_i(x) = \omega(\phi_i(x))$$

for every non-zero element $x \in A_i$, and this definition does not depend on the choice of ϕ_i. Indeed, we can reformulate it in the more intrinsic way

$$\omega_i(x) = m \quad \text{if and only if} \quad x \in \mathfrak{n}_i^m \setminus \mathfrak{n}_i^{m+1}.$$

In particular, $\omega_i(x) > 0$ if $x \in \mathfrak{m}_i$, and we have the positive integer

$$e_i = \min\{\omega_i(x) \,|\, x \in \mathfrak{m}_i\}.$$

Using these notations we will express multiplicities in a different way.

Proposition 3.3 *It holds:*

a) $\deg(\varphi) = \omega_i(x)f_i \geq e_i f_i$ *for any non-trivial analytic homomorphism $\varphi : \mathbb{K}\{\mathbf{t}\} \to A_i$ with $\varphi(\mathbf{t}) = x$.*

b) $\mu_i(A) = e_i f_i$.

3 Multiplicity in Dimension 1

c) $\mu(A) = \sum_{i=1}^{s} e_i f_i$.

Proof. a) The homomorphism φ induces another one $\psi : \mathbb{K}\{t\} \to A_i \subset B_i \simeq \mathbb{K}'_i\{t\}$ and we have to show that the degree d of ψ is

$$d = \omega_i(x)[\mathbb{K}'_i : \mathbb{K}].$$

Indeed, set $p = \omega_i(x) = \omega(\psi(t))$. After an isomorphism of the target we can assume $\psi(t) = \pm t^p$ (II.4.7 a)). We thus get the factorization

$$\mathbb{K}\{t\} \to \mathbb{K}\{t\} \to \mathbb{K}'_i\{t\},$$

where the first arrow is $t \mapsto \pm t^p$ and the second one the canonical inclusion.

The first homomorphism gives a field extension of degree p. Indeed, any power series $f \in \mathbb{K}\{t\}$ can be written as follows

$$f = (a_0 + a_1 t + \cdots + a_{p-1} t^{p-1}) + (a_p t^p + a_{p+1} t^{p+1} + \cdots + a_{2p-1} t^{2p-1}) - \cdots =$$
$$= (a_0 + a_p t^p + \cdots) + (a_1 + a_{p+1} t^p + \cdots)t + \cdots + (a_{p-1} + a_{2p-1} t^p + \cdots)t^{p-1} =$$
$$= f_0(t^p) + f_1(t^p)t + \cdots + f_{p-1}(t^p)t^{p-1},$$

which shows that $1, t, \ldots, t^{p-1}$ are a basis of that first field extension.

As the degree of the second extension is clearly $[\mathbb{K}'_i : \mathbb{K}]$, we have proved that the degree of the extension corresponding to ψ is exactly $p[\mathbb{K}'_i : \mathbb{K}]$, as claimed.

On the other hand since $x \in \mathfrak{m}_i$, by the definition of e_i, it is $\omega(x) \geq e_i$, and so the proof of *a)* is complete.

b) From *a)* we deduce one inequality, namely $\mu_i(A) \geq e_i f_i$. To prove the converse one we have to find some φ such that

$$\omega_i(\varphi(t)) = e_i.$$

In order to do so we apply the construction II.2.3: choose any generators $x_1, \ldots, x_n$ of $\mathfrak{m}_i$ and consider the surjective analytic homomorphism $\mathcal{O}_n \to A_i$ which maps $\mathbf{x}_j$ to x_j, $1 \leq j \leq n$. Then

a) $\varphi(t) = x_1$.

b) *There are $n-1$ polynomials $P_\ell \in \mathcal{O}_{n-\ell}[\mathbf{x}_{n-\ell+1}]$, $1 \leq \ell \leq n-1$, such that*

$$P_\ell(x_1, \ldots, x_{n-\ell}, x_{n-\ell+1}) = 0.$$

c) *Every P_ℓ, $1 \leq \ell \leq n-1$, is a monic polynomial whose degree equals its order as a power series.*

We claim that this is the homomorphism we seek.

We first show that

$$\omega_i(\varphi(t)) = \omega_i(x_1) \leq \omega_i(x_j)$$

for all j. We do this by induction, the assertion being trivial for $j = 1$. To check it for $j > 1$ consider the polynomial P_{n-j+1}. By the properties above, we get a monic equation
$$x_j^q + a_1 x_j^{q-1} + \cdots + a_q = 0,$$
where the coefficients verify
$$a_k \in (x_1, \ldots, x_{j-1})^k, \ 1 \leq k \leq q.$$
Thus, by the properties of ω_i and the induction hypothesis, we get
$$\omega_i(a_k) \geq \min_{|\nu|=k} \omega_i(x_1^{\nu_1} \cdots x_{j-1}^{\nu_{j-1}}) \geq k\omega_i(x_1)$$
for $1 \leq k \leq q$. Hence, if $\omega_i(x_j) < \omega_i(x_1)$ we would get
$$\omega_i(a_k x_j^{q-k}) \geq k\omega_i(x_1) + (q-k)\omega_i(x_j) > q\omega_i(x_j) = \omega_i(x_j^q),$$
and so
$$\omega_i(x_j^q) < \omega(a_1 x_j^{q-1} + \cdots + a_q),$$
which is impossible, since $a_1 x_j^{q-1} + \cdots + a_q = -x_j^q$.

Next, the maximal ideal $\mathfrak{m}_i$ of A_i is generated by $x_1, \ldots, x_n$ of $\mathfrak{m}$. Hence any $x \in \mathfrak{m}_i$ can we written as
$$x = a_1 x_1 + \cdots + a_n x_n, \ a_1, \ldots, a_n \in A_i,$$
and by the properties of ω we get
$$\omega_i(x) \geq \min\{\omega_i(x_1), \ldots, \omega_i(x_n)\} = \omega_i(x_1).$$
Thus $\omega_i(x_1) = e_i$, and we have finished the proof of $b)$.

Finally, $c)$ is an immediate consequence of $b)$ and the definition of multiplicity. □

Remarks 3.4 *a)* We have actually shown that in order to compute multiplicities the changes of coordinates described in II.2.3 are the good ones. This has the further consequence that a generic choice of coordinates (see Remark II.2.6) gives the multiplicity of A.

b) Suppose that A is *planar*, that is, $A = \mathbb{K}\{\mathbf{x}, \mathbf{y}\}/(h)$. Then $\mu(A) = \omega(h)$.

Indeed, after a linear change of coordinates we may assume h is regular of order $p = \omega(h)$ (Lemma I.3.1), and then $h = uP$ where u is a unit and $P \in \mathbb{K}\{\mathbf{x}\}[\mathbf{y}]$ a distinguished polynomial of degree p (Weierstrass's Preparation Theorem). We then have a factorization $P = P_1 \cdots P_s$, into distinguished polynomials, which are irreducible both as polynomials and as power series (Remark II.2.2). Note that this factorization has no multiple factor because A is reduced, and consequently the branches of A are $A_i = \mathbb{K}\{\mathbf{x}, \mathbf{y}\}/(P_i)$, $1 \leq i \leq s$. Let p_i denote the degree of P_i. Clearly $\omega(P_i) \leq p_i$ and

$$p = \sum_i p_i \geq \sum_i \omega(P_i) = \omega(P) = p.$$

Hence $\omega(P_i) = p_i$, and by *a)* the coordinates are good to compute multiplicities: $\mu_i(A)$ is the degree of the field extension corresponding to $\mathbb{K}\{\mathbf{x}\} \to \mathbb{K}\{\mathbf{x},\mathbf{y}\}/(P_i)$. But the canonical homomorphism

$$\mathbb{K}\{\mathbf{x}\}[\mathbf{y}]/(P_i) \to \mathbb{K}\{\mathbf{x},\mathbf{y}\}/(P_i)$$

is a bijection by Rückert's Division Theorem, and consequently that field extension is $\mathbb{K}(\{\mathbf{x}\}) \subset \mathbb{K}(\{\mathbf{x}\})[\mathbf{y}]/(P_i)$. As the degree of this extension is the degree p_i of P_i, we have

$$\mu(A) = \sum_i \mu_i(A) = \sum_i p_i = p = \omega(h).$$

□

c) An easy consequence of Propositions 2.7 and 3.3 is that the multiplicities of an analytic ring over $\mathbb{R}$ and its complexification coincide. We leave the proof as an exercise (note that $f_i \neq 1$ if and only if the branch A_i is not 2-real).

We end this section with a classical construction.

(3.5) Quadratic transforms. We keep all the notations and data already introduced. Let $x \in \mathfrak{m}_i$ have value $\omega_i(x) = e_i$. Then

a) $A_i^{(1)} = A_i[x^{-1}\mathfrak{m}_i] \subset B_i$.

Indeed, if $y \in \mathfrak{m}_i$, then $\omega_i(y) \geq e_i = \omega_i(x)$, or equivalently $\omega(\phi_i(y)) \geq \omega(\phi_i(x))$. Thus $\phi_i(y)/\phi_i(x)$ is a well defined power series in $\mathbb{K}'_i\{t\}$, and since ϕ_i is an isomorphism, $y/x \in B_i$. □

b) $A_i^{(1)}$ *is an analytic ring of dimension 1 whose maximal ideal is* $\mathfrak{m}_i^{(1)} = \mathfrak{n}_i \cap A_i^{(1)}$.

In fact, as B_i is a finite A_i-module, $A_i^{(1)} \subset B_i$ is a finite A_i-module, too, and the first assertion follows from Lemma 1.1. Now $\mathfrak{n}_i \cap A_i^{(1)}$ is a prime ideal of $A_i^{(1)}$, obviously $\neq (0)$, and consequently it is the maximal ideal. □

c) $A_i^{(1)}$ *does not depend on the choice of* x.

To see this, let $y \in A$ have also value $\omega_i(y) = e_i$. Then $\omega_i(y/x) = 0$ and arguing as above it follows that y/x is a unit of B_i. As $y/x \in A_i^{(1)}$, and $\mathfrak{m}_i^{(1)} = \mathfrak{n}_i \cap A_i^{(1)}$, we conclude that y/x is a unit in $A_i^{(1)}$. Hence, $x/y \in A_i^{(1)}$, and so

$$y^{-1}\mathfrak{m}_i = (x/y)(x^{-1}\mathfrak{m}_i) \subset A_i^{(1)}.$$

Whence $A[y^{-1}\mathfrak{m}_i] \subset A[x^{-1}\mathfrak{m}_i]$. The other inclusion follows by symmetry, and we conclude $A[y^{-1}\mathfrak{m}_i] = A[x^{-1}\mathfrak{m}_i]$. □

d) $A_i = A_i^{(1)}$ *if and only if* $A_i = B_i$.

Let $x_1, \ldots, x_n \in A_i$ generate $\mathfrak{m}_i$. Then
$$x^{-1}x_1, \ldots, x^{-1}x_n \in A_i^{(1)} = A_i$$
implies $\mathfrak{m}_i = xA_i$. Thus $\mathfrak{m}_i$ is principal, and it follows that A_i is a normal domain, so that $A_i = B_i$. □

Clearly, if $A_i \neq A_i^{(1)}$, we can repeat the construction to get a sequence of analytic rings of dimension 1
$$A = A_i^{(0)} \subset A_i^{(1)} \subset \cdots \subset A_i^{(\ell)} \subset \cdots \subset B_i.$$

By d), if $A_i^{(r)} = A_i^{(r+1)}$, then $A_i^{(r)} = B_i$, and the sequence stops. But this actually happens, because B_i is a finite A_i-module, and consequently a noetherian A_i-module. Thus we come to the

Definition 3.6 *In the situation above, the ring* $A_i^{(\ell)}$ *is called the ℓ-th quadratic transform of* A_i, *and the sequence of analytic rings*
$$A_i = A_i^{(0)} \subset A_i^{(1)} \subset \cdots \subset A_i^{(r)} = B_i$$
is called the sequence of quadratic transforms of A_i.

Again we leave as an exercise to check that *the complexification of a sequence of quadratic transforms is a sequence of quadratic transforms of the complexification.*

4 Newton-Puiseux's Theorem

We start by defining power series with rational exponents over $\mathbb{K} = \mathbb{R}$ or $\mathbb{C}$.

Definitions and Notations 4.1 A *formal Puiseux series* in the indeterminate $\mathbf{t}$ is an expression $f = \sum_{m \geq 0} a_m \mathbf{t}^{m/p}$, in short $\sum_m a_m \mathbf{t}^{m/p}$ or $\sum a_m \mathbf{t}^{m/p}$, where $a_m \in \mathbb{K}$ for every m and p is an integer ≥ 1. The a_m's are the *coefficients* of f, and the first of them a_0 is denoted by $f(0)$.

Consider now two formal Puiseux series $f = \sum a_m \mathbf{t}^{m/p}$ and $g = \sum b_m \mathbf{t}^{m/q}$. If $p = q$ we define $f + g$ and fg as we did for ordinary formal power series (I.2.4). Otherwise, if $p \neq q$, we first write
$$f = \sum a_m \mathbf{t}^{m/p} = \sum a_m \mathbf{t}^{mq/pq}; \quad g = \sum b_m \mathbf{t}^{m/q} = \sum b_m \mathbf{t}^{mp/qp},$$
and then sum and multiply as said before. Thus the set $\mathbb{K}[[\mathbf{t}^*]]$ of all formal Puiseux series becomes a ring.

For every $p \geq 1$ let $\mathbb{K}[[\mathbf{t}^{1/p}]]$ be the subring of $\mathbb{K}[[\mathbf{t}^*]]$ consisting of all Puiseux series of the form $\sum a_m \mathbf{t}^{m/p}$. We have:
$$\mathbb{K}[[\mathbf{t}^*]] = \bigcup_{p \geq 1} \mathbb{K}[[\mathbf{t}^{1/p}]].$$

4 Newton-Puiseux's Theorem

Note that for $p = 1$ in this union we just obtain the ring $\mathbb{K}[[t]]$ of ordinary formal power series.

It is also easy to check that $\mathbb{K}[[t^*]]$ is a domain, and its quotient field $\mathbb{K}((t^*))$ can be described as

$$\mathbb{K}((t^*)) = \bigcup_{p \geq 1} \mathbb{K}((t^{1/p})),$$

where of course $\mathbb{K}((t^{1/p}))$ stands for the quotient field of $\mathbb{K}[[t^{1/p}]]$.

Let now $\mathbf{x}$ be another indeterminate. For every $p \geq 1$ we have a canonical isomorphism of $\mathbb{K}$-algebras

$$\tau_p : \mathbb{K}[[\mathbf{x}]] \to \mathbb{K}[[t^{1/p}]] : h \mapsto h(t^{1/p}),$$

defined quite obviously by *substitution by* $t^{1/p}$. Using these τ_p's and the formulas above any problem concerning finitely many Puiseux series can be reduced to a problem concerning ordinary power series.

The τ_p's also give an easy way to introduce convergence. A formal Puiseux series $f = \sum a_m t^{m/p}$ is called *convergent* if the ordinary power series $\tau_p^{-1}(f) = \sum a_m \mathbf{x}^m$ is convergent. All the convergent Puiseux series form a subring $\mathbb{K}\{t^*\}$ of the ring $\mathbb{K}[[t^*]]$, whose quotient field will be denoted by $\mathbb{K}(\{t^*\})$. We have formulas similar to the ones above:

$$\mathbb{K}\{t^*\} = \bigcup_{p \geq 1} \mathbb{K}\{t^{1/p}\}, \quad \mathbb{K}(\{t^*\}) = \bigcup_{p \geq 1} \mathbb{K}(\{t^{1/p}\}).$$

Again, for $p = 1$ we obtain ordinary convergent power series.

Proposition 4.2 *We have:*

a) *Every* $f \in \mathbb{K}\{t^*\}$ *can be uniquely written as* $f = t^{m/p} u$, *for some rational number* $m/p \geq 0$ *and some unit* u *of* $\mathbb{K}\{t^*\}$.

b) *The set of all* $f \in \mathbb{K}\{t^*\}$ *with* $f(0) = 0$ *is the unique prime ideal* $\mathfrak{m}^* \neq (0)$ *of* $\mathbb{K}\{t^*\}$.

c) $\mathbb{K}\{t^*\}$ *is integrally closed in its quotient field.*

d) $\mathbb{K}\{t^*\}$ *is integral over* $\mathbb{K}\{t\}$.

The same result holds true replacing $\mathbb{K}\{t^*\}$ *by* $\mathbb{K}[[t^*]]$ *and* $\mathbb{K}\{t\}$ *by* $\mathbb{K}[[t]]$.

Proof. a) There is some p such that $f \in \mathbb{K}\{t^{1/p}\}$. Since the assertion is true in $\mathbb{K}\{\mathbf{x}\}$, we can apply τ_p to get a unit u such that $f = t^{m/p} u$. Given then a second expression $f = t^{n/q} v$, we have both in $\mathbb{K}\{t^{1/pq}\}$, and using τ_{pq} we can work in $\mathbb{K}\{\mathbf{x}\}$. But uniqueness holds in $\mathbb{K}\{\mathbf{x}\}$, and we are done.

b) Clearly the condition of the statement defines an ideal $\mathfrak{m}^*$. Furthermore, if $f(0) \neq 0$, in the expression $f = t^{m/p} u$ of *a)* we must have $m = 0$. Thus $f = u$ is a unit. This shows that $\mathfrak{m}^*$ is the unique maximal ideal of $\mathbb{K}\{t^*\}$.

Now let $\mathfrak{p} \neq (0)$ be a prime ideal of $\mathbb{K}\{\mathbf{t}^*\}$. We can choose $f \in \mathfrak{p}$, $f \neq 0$, and from a) write $f = \mathbf{t}^{m/p} u$ for some unit u. Then
$$\mathbf{t}^m = (u^{-1}f)^p \in \mathfrak{p},$$
and since $\mathfrak{p}$ is prime we conclude $\mathbf{t} \in \mathfrak{p}$. On the other hand any $g \in \mathfrak{m}^*$ can be written as $g = \mathbf{t}^{n/q} v$ with $n > 0$, and so
$$g^q = \mathbf{t}^n v^q = (\mathbf{t}^{n-1} v^q)\mathbf{t} \in \mathfrak{p}.$$
As $\mathfrak{p}$ is prime, $g \in \mathfrak{p}$. Thus $\mathfrak{m}^* \subset \mathfrak{p}$ and the equality follows because $\mathfrak{m}^*$ is maximal.

c) Let $h \in \mathbb{K}(\{\mathbf{t}^*\})$ be integral over $\mathbb{K}\{\mathbf{t}^*\}$, that is
$$h^s + a_1 h^{s-1} + \cdots + a_s = 0, \quad a_1, \ldots, a_s \in \mathbb{K}\{\mathbf{t}^*\}.$$
For a suitable $p \geq 1$, $h \in \mathbb{K}(\{\mathbf{t}^{1/p}\})$ and $a_k \in \mathbb{K}\{\mathbf{t}^{1/p}\}$ for all k. We can thus apply the isomorphism τ_p to translate the above equation to $\mathbb{K}\{\mathbf{x}\}$, which is integrally closed in its quotient field. We conclude that $\tau_p^{-1}(h) \in \mathbb{K}\{\mathbf{x}\}$, and $h \in \mathbb{K}\{\mathbf{t}^{1/p}\}$.

d) We have to see that every element of $\mathbb{K}\{\mathbf{t}^*\}$ verifies a monic equation with coefficients in $\mathbb{K}\{\mathbf{t}\}$. Consider, thus, $p > 0$, $f \in \mathbb{K}\{\mathbf{x}\}$ and $f(\mathbf{t}^{1/p}) \in \mathbb{K}\{\mathbf{t}^*\}$. We first look for a monic polynomial $P(\mathbf{z}, \mathbf{y}) \in \mathbb{K}\{\mathbf{z}\}[\mathbf{y}]$ such that $P(\mathbf{x}^p, f(\mathbf{x})) = 0$. To find it, we take the analytic homomorphism $\mathbb{K}\{\mathbf{z}, \mathbf{y}\} \to \mathbb{K}\{\mathbf{x}\}$ defined by
$$\mathbf{z} \mapsto \mathbf{x}^p, \quad \mathbf{y} \mapsto f(\mathbf{x}).$$
This homomorphism is finite (Example II.1.6), and consequently cannot be injective, (an injective finite homomorphism $\mathbb{K}\{\mathbf{z}, \mathbf{y}\} \to \mathbb{K}\{\mathbf{x}\}$ would give $\dim(\mathbb{K}\{\mathbf{x}\}) = \dim(\mathbb{K}\{\mathbf{z}, \mathbf{y}\}) = 2$). Hence there is some series $H \in \mathbb{K}\{\mathbf{z}, \mathbf{y}\}$ such that $H(\mathbf{x}^p, f(\mathbf{x})) = 0$. We write $H = \mathbf{z}^n Q$ with $Q(0, \mathbf{y}) \neq 0$ and by Weierstrass's Preparation Theorem (Proposition I.3.3), there are a distinguished polynomial $P \in \mathbb{K}\{\mathbf{z}\}[\mathbf{y}]$ and a unit $u \in \mathbb{K}\{\mathbf{z}, \mathbf{y}\}$ with $Q = uP$. Hence
$$0 = H(\mathbf{x}^p, f(\mathbf{x})) = \mathbf{x}^{pn} u(\mathbf{x}^p, f(\mathbf{x})) Q(\mathbf{x}^p, f(\mathbf{x})).$$
As $u(0,0) \neq 0$, $u(\mathbf{x}^p, f(\mathbf{x})) \neq 0$, and $P(\mathbf{x}^p, f(\mathbf{x})) = 0$. We thus have found the polynomial we sought, and after the substitution $x \mapsto \mathbf{t}^{1/p}$ it gives us an equation of integral dependence of $f(\mathbf{t}^{1/p})$ over $\mathbb{K}\{\mathbf{t}\}$. $\square$

We are now ready to state Newton-Puiseux's Theorem, but first we consider complexification in this new context:

Lemma 4.3 *The ring $\mathbb{R}\{\mathbf{t}^*\}$ is 2-real, and $\mathbb{R}\{\mathbf{t}^*\}[\sqrt{-1}] = \mathbb{C}\{\mathbf{t}^*\}$. The same holds in the formal case.*

Proof. Suppose that $f^2 + g^2 = 0$ with $f, g \in \mathbb{R}\{\mathbf{t}^*\}$ and $f \neq 0$. Then $f, g \in \mathbb{R}\{\mathbf{t}^{1/p}\}$ for a suitable p. and via the isomorphism τ_p we would conclude that $\mathbb{R}\{\mathbf{x}\}$ is not 2-real, which is false. Consider now a series $h(\mathbf{t}^{1/p}) \in \mathbb{C}\{\mathbf{t}^*\}$ with $h \in \mathbb{C}\{\mathbf{x}\}$. We write $h = f + \sqrt{-1} g$ with $f, g \in \mathbb{R}\{\mathbf{x}\}$ and get $h(t^{1/p}) = f(\mathbf{t}^{1/p}) + \sqrt{-1} g(\mathbf{t}^{1/p})$. $\square$

4 Newton-Puiseux's Theorem

Proposition 4.4 *(Newton-Puiseux's Theorem)* We have:

a) *The field $\mathbb{C}(\{t^*\})$ (resp. $\mathbb{C}((t^*)))$ is algebraically closed.*

b) *The field $\mathbb{R}(\{t^*\})$ (resp. $\mathbb{R}((t^*)))$ is real closed.*

Proof. By Lemma 4.4 $\sqrt{-1} \notin \mathbb{R}(\{t^*\})$, and $\mathbb{C}(\{t^*\}) = \mathbb{R}(\{t^*\})[\sqrt{-1}]$. These two conditions imply that the two assertions in the statement are equivalent (general theory of formally real fields [L XI.2]) Hence we will only prove a).

We have to see that every polynomial $P \in \mathbb{C}(\{t^*\})[y]$ of degree $p \geq 1$ has some root. Multiplying by a common denominator of the coefficients we may assume that $P \in \mathbb{C}\{t^*\}[y]$. Then, by considering the polynomial

$$a^{p-1}P(y/a),$$

where a is the coefficient of the monomial of maximal degree p, we can suppose that P is monic. We have hence

$$P = y^p + a_1(t^{1/q})y^{p-1} + \cdots + a_p(t^{1/q}),$$

with $a_1, \ldots, a_p \in \mathbb{C}\{x\}$. We set now

$$P^* = y^p + a_1(x)y^{p-1} + \cdots + a_p(x) \in \mathbb{C}\{x, y\}.$$

Let $c \in \mathbb{C}$ be a root of multiplicity, say, $q \geq 1$ of the polynomial $P^*(0, y) \in \mathbb{C}[y]$. After the change $y = y + c$ we are reduced to the case $c = 0$, and from Hensel's Lemma (Proposition I.3.4) we get a factorization $P^* = QQ'$, where Q, Q' are monic polynomials of $\mathbb{C}\{x\}[y]$, Q has degree q and $Q(0, y) = y^q$. In particular, Q is a distinguished polynomial. By Remark II.2.2 we can factorize Q into irreducible distinguished polynomials $Q_1, \ldots, Q_s$ which are irreducible as series in $\mathbb{C}\{x, y\}$. Pick Q_1 for instance, and put $A = \mathbb{C}\{x, y\}/Q_1$. This analytic ring is a domain of dimension 1, because Q_1 is irreducible, and its normalization is isomorphic to $\mathbb{C}\{z\}$ (Proposition 3.2). Thus we get an injective analytic homomorphism $A \to \mathbb{C}\{z\}$, and from this another

$$\varphi : \mathbb{C}\{x, y\} \to \mathbb{C}\{z\}$$

whose kernel is $Q_1\mathbb{C}\{x, y\}$. Then $\varphi(x) \neq 0$, since otherwise x would belong to $Q_1\mathbb{C}\{x, y\}$ and Q_1 would divide x. Thus $\varphi(x) \in \mathbb{C}\{z\}$ is a series of order $m > 0$, and by II.4.7 we can compose φ with an automorphism of $\mathbb{C}\{z\}$ to get

$$\varphi(x) = z^m.$$

Finally, let $\varphi(y) = f(z)$, and so

$$Q_1(z^m, f(z)) = \varphi(Q_1) = 0.$$

Since Q_1 is a factor of P^* in $\mathbb{C}\{x\}[y]$, we also get

$$P^*(z^m, f(z)) = 0.$$

This means that $f(t^{1/m}) \in \mathbb{C}\{t^*\}$ is a root of P and the proof is finished. □

The preceding proof can be examined more carefully to get further interesting information:

Proposition 4.5 *Let $P \in \mathbb{C}\{t\}[y]$ be an irreducible distinguished polynomial. Then:*

  a) *The degree of P is the smallest integer p such that P has a root of the form $h = f(t^{1/p})$ for some $f \in \mathbb{C}\{x\}$.*

  b) *Any other root of P is then of the form $f(\xi t^{1/p})$ for some p-th root of unity $\xi \in \mathbb{C}$.*

Proof. a) Set $A = \mathbb{C}\{t, y\}/(P)$. This is an analytic ring of dimension 1 over $\mathbb{C}$ and, since P is irreducible, it is a domain. The normalization B of A is then isomorphic to $\mathbb{C}\{x\}$, and after an automorphism of $\mathbb{C}\{x\}$ we may assume $t = x^p$. If $y = f(x)$, we get $P(x^p, f(x)) = 0$ and $h = f(t^{1/p})$ is a root of P. Once we have found this root, we will prove that p is the smallest integer considered in $a)$, and that it verifies $b)$.

The root h comes from the homomorphism

$$\varphi : \mathbb{C}\{t, y\}/(P) = A \to B \simeq \mathbb{C}\{x\} : t \mapsto x^p, \; y \mapsto f(x),$$

and conversely, any root $g(t^{1/q})$ of P defines

$$\phi : \mathbb{C}\{t, y\}/(P) = A \to \mathbb{C}\{z\} : t \mapsto z^q, \; y \mapsto g(z).$$

We claim that there is a third homomorphism $\psi : B \simeq \mathbb{C}\{x\} \to \mathbb{C}\{z\}$ such that $\psi \circ \varphi = \phi$.

Indeed, since ϕ is injective, it extends to the quotient field of A, and since B is contained in that quotient field we get $\psi : B \simeq \mathbb{C}\{x\} \to \mathbb{C}(\{z\})$. But B is integral over A, and so $\psi(B)$ is integral over $\psi(A) = \phi(A) \subset \mathbb{C}\{z\}$. As $\mathbb{C}\{z\}$ is normal, we conclude $\psi(B) \subset \mathbb{C}\{z\}$.

Once ψ is available, we can write

$$\psi(x)^p = \psi(x^p) = \psi\varphi(t) = \phi(t) = z^q.$$

It follows that $p \,|\, q$, and that $\psi(x)$ is a p-th root of z^q:

$$\psi(x) = \xi z^{q/p}$$

for some p-th root of unity $\xi \in \mathbb{C}$. Thus we get

$$g(z) = \phi(y) = \psi\varphi(y) = \psi(f(x)) = f(\psi(x)) = f(\xi z^{q/p})$$

and so $g(t^{1/q}) = f(\xi t^{1/p})$.

4 Newton-Puiseux's Theorem

After all of this, it only remains to show that p is the degree of P. For that, note that since P is irreducible, it has no multiple root, and, since $\mathbb{C}(\{\mathbf{t}^*\})$ is algebraically closed (Newton-Puiseux's Theorem, Proposition 4.4), we conclude that the degree of P is the number of its roots. Hence we need to see that when ξ runs among the p-th roots of unity, the Puiseux series

$$g(\mathbf{t}^{1/q}) = f(\xi \mathbf{t}^{1/p})$$

are all different. But suppose

$$f(\xi \mathbf{t}^{1/p}) = f(\xi' \mathbf{t}^{1/p}).$$

Then $f(\xi \mathbf{x}) = f(\xi' \mathbf{x})$, and after the substitution $\mathbf{x} = \mathbf{x}/\xi$ we get $f(\mathbf{x}) = f(\zeta \mathbf{x})$, where $\zeta = \xi'/\xi$ is another p-th root of unity. This means that if $f = \sum_k a_k \mathbf{x}^k$, it holds $a_k = \zeta^k a_k$ for all k, or in other words,

$$\zeta^k = 1 \text{ if } a_k \neq 0.$$

Now, let r be the smallest positive integer such that $\zeta^r = 1$; as is well known, $r \mid p$, say $rs = p$. We then write $k = mr + \rho(k)$ with $0 \leq \rho(k) < r$, and the last condition reads

$$\zeta^{\rho(k)} = 1 \text{ if } a_k \neq 0.$$

By the choice of r, $\zeta^{\rho(k)} = 1$ if and only if $\rho(k) = 0$, if and only if $r \mid k$. Hence $a_k = 0$ if r does not divide k, that is, $f = g(\mathbf{x}^r)$ for a series $g \in \mathbb{C}\{\mathbf{x}\}$. We conclude

$$f(\mathbf{t}^{1/p}) = g(\mathbf{t}^{r/p}) = g(\mathbf{t}^{1/s}),$$

and $s = p$ by the minimality of p. Whence $\zeta = 1$ and $\xi = \xi'$, which finishes the proof. $\square$

Remarks 4.6 The proof of Proposition 4.4 tells also how to treat an arbitrary monic polynomial $P \in \mathbb{C}\{\mathbf{t}\}[\mathbf{y}]$. Namely, if P is irreducible it follows from Hensel's Lemma (Proposition I.3.4) that $P(0, \mathbf{y}) = (\mathbf{y} - c)^p$ for some $c \in \mathbb{C}$. Consequently after the substitution $\mathbf{y} = \mathbf{y} + c$ we obtain a distinguished polynomial and can apply Proposition 4.5; if P is not irreducible, we first split it into irreducible factors. In any case, if p is the degree of P, all roots of P are in $\mathbb{C}\{\mathbf{t}^{1/p!}\}$.

IV Nullstellensätze

Summary. This chapter is devoted to the real and complex Nullstellensätze. In the complex case the Nullstellensatz is a direct consequence of Rückert's Parametrization. In the real case two other results are essential, the Homomorphism Theorem and the solution to Hilbert's 17th Problem. We consider only the analytic case, remarking that everything can be done analogously in the formal one.

1 Zero Sets and Zero Ideals

According to Definitions and Notations III.4.1, let $\mathbb{K}\{\mathbf{t}^*\}$ be the ring of convergent Puiseux series over $\mathbb{K} = \mathbb{R}$ or $\mathbb{C}$, $\mathbb{K}(\{\mathbf{t}^*\})$ its quotient field and $\mathfrak{m}^*$ its maximal ideal. We will also use the notations

$$W = \mathbb{K}\{\mathbf{t}^*\}, \quad U = \mathfrak{m}^*, \quad F = \mathbb{K}(\{\mathbf{t}^*\}).$$

(1.1) Fix an integer $n \geq 0$ and put $D = U \times \cdots \times U \subset F^n$. Let $f \in \mathcal{O}_n$. Then for every $x(\mathbf{t}) = (x_1(\mathbf{t}), \ldots, x_n(\mathbf{t})) \in D$ the substitution $f(x(\mathbf{t}))$ is a well defined convergent Puiseux series. Indeed, up to an isomorphism τ_p (III.4.1) this is only a substitution of ordinary convergent power series. An *associated function* can thus be defined by

$$^a f : D = U \times \cdots \times U \to W \subset F : x(\mathbf{t}) \mapsto f(x(\mathbf{t})).$$

An intuitive way to see this construction is the following. The field F is an extension of $\mathbb{K}$ which contains an *infinitesimal* $\mathbf{t}$, that is, an element which belongs to every neighborhood of 0 in $\mathbb{K}$. Then D is a neighborhood of the origin in F^n which is small enough to be contained in the convergence domain $D(f)$ of any series $f \in \mathcal{O}_n$.

Definition 1.2 *Let I be an ideal of $\mathcal{O}_n$. The* zero set *of I is the set*

$$\mathcal{Z}(I) = \{x(\mathbf{t}) \in D \mid f(x(\mathbf{t})) = 0 \text{ for all } f \in I\}.$$

The familiar properties of this operator are:

Proposition 1.3 *Let I, J be two ideals of $\mathcal{O}_n$. Then:*

a) $\mathcal{Z}(I) \subset \mathcal{Z}(J)$ *if* $I \supset J$.

b) $\mathcal{Z}(I \cdot J) = \mathcal{Z}(I \cap J) = \mathcal{Z}(I) \cup \mathcal{Z}(J)$.

c) $\mathcal{Z}(I + J) = \mathcal{Z}(I) \cap \mathcal{Z}(J)$.

1 Zero Sets and Zero Ideals

Proof. *a)* is immediate. For *b)* note that *a)* implies the inclusions

$$\mathcal{Z}(I \cdot J) \supset \mathcal{Z}(I \cap J) \supset \mathcal{Z}(I) \cup \mathcal{Z}(J).$$

Now let $x(\mathbf{t}) \in \mathcal{Z}(I \cdot J) \setminus \mathcal{Z}(I)$, that is, $f(x(\mathbf{t})) \neq 0$ for some $f \in I$. Then, for every $g \in J$ we get

$$0 = (gf)(x(\mathbf{t})) = g(x(\mathbf{t}))f(x(\mathbf{t})),$$

and so $g(x(\mathbf{t})) = 0$. Thus $x(\mathbf{t}) \in \mathcal{Z}(J)$. We leave the last formula *c)* as an exercise. □

The other standard operator is:

Definition 1.4 *Let $Y \subset D$. The* zero ideal *of Y is the ideal*

$$\mathcal{J}(Y) = \{f \in \mathcal{O}_n \mid f(x(\mathbf{t})) = 0 \text{ for all } x(\mathbf{t}) \in Y\}.$$

It is clear that the zero ideal is indeed an ideal, and the following properties are also immediate:

Proposition 1.5 *Let Y, Z be two subsets of D. Then:*

a) $\mathcal{J}(Y) \subset \mathcal{J}(Z)$ *if* $Y \supset Z$.

b) $\mathcal{J}(Y \cup Z) = \mathcal{J}(Y) \cap \mathcal{J}(Z)$.

The problem solved by the Nullstellensätze is the determination of the ideal $\mathcal{J}(\mathcal{Z}(I))$ for any given ideal I. There are two different solutions according to whether $\mathbb{C}$ or $\mathbb{R}$ is considered as coefficient field $\mathbb{K}$, as we will find in the coming sections. First we introduce an equivalent description of zero sets that will be useful later.

Proposition 1.6 *Let Φ denote the collection of all homomorphisms of $\mathbb{K}$-algebras $\mathcal{O}_n \to \mathbb{K}\{\mathbf{t}^*\}$, and set*

$$\Phi(I) = \{\varphi \in \Phi \mid \ker(\varphi) \supset I\}$$

for $I \subset \mathcal{O}_n$. Then we have:

a) Every homomorphism $\varphi \in \Phi$ is local and it is defined by the substitution

$$\varphi(f) = f(x_1(\mathbf{t}), \ldots, x_n(\mathbf{t})), \ f \in \mathcal{O}_n,$$

where $x_i(\mathbf{t}) = \varphi(\mathbf{x}_i) \in \mathfrak{m}^*$ for $1 \leq i \leq n$.

b) The correspondence: $\varphi \mapsto (\varphi(\mathbf{x}_1), \ldots, \varphi(\mathbf{x}_n))$ is a bijection from Φ onto the set $D = U \times \cdots \times U \subset F^n$.

c) This bijection maps $\Phi(I)$ onto the zero set $\mathcal{Z}(I)$ for every ideal $I \subset \mathcal{O}_n$.

d) $\mathcal{J}(\mathcal{Z}(I)) = \bigcap_{\varphi \in \Phi(I)} \ker(\varphi)$.

Proof. a) To see that φ is local, we refer to the proof that every analytic homomorphism is local (Proposition II.1.3). Once this is known, we set

$$\varphi(\mathbf{x}_i) = x_i(\mathbf{t}) \in \mathfrak{m}^*$$

for $1 \le i \le n$, and define $\phi \in \Phi$ by

$$\phi(f) = f(x_1(\mathbf{t}), \ldots, x_n(\mathbf{t})), \quad f \in \mathcal{O}_n.$$

We will show that $\varphi = \phi$ using a proof similar to that of Proposition II.1.3 a), but here the argument ends differently since Krull's Theorem fails for the ring $\mathbb{K}\{\mathbf{t}^*\}$. First, we put

$$x_i(\mathbf{t}) = u_i \mathbf{t}^{m_i/p_i},$$

where u_i is a unit, for $1 \le i \le n$, and

$$\frac{m_0}{p_0} = \min\left\{\frac{m_1}{p_1}, \ldots, \frac{m_n}{p_n}\right\}.$$

Suppose now that there is an $f \in \mathcal{O}_n$ such that $\varphi(f) \ne \phi(f)$, so that we can write

$$\varphi(f) - \phi(f) = u\mathbf{t}^{m/p},$$

where u is a unit; we then choose an integer s such that $s\frac{m_0}{p_0} \ge 1 + \frac{m}{p}$, and set

$$f = g + \sum_{|\nu|=s} h_\nu \mathbf{x}_1^{\nu_1} \cdots \mathbf{x}_n^{\nu_n},$$

with $g \in \mathbb{K}[\mathbf{x}_1, \ldots, \mathbf{x}_n]$ and $h_\nu \in \mathcal{O}_n$ for $|\nu| = s$. Since φ and ϕ are homomorphisms of $\mathbb{K}$-algebras, we obtain

$$u\mathbf{t}^{\frac{m}{p}} = \varphi(f) - \phi(f) =$$

$$= \sum_{|\nu|=s} (\varphi(h_\nu) - \phi(h_\nu))\, x_1(\mathbf{t})^{\nu_1} \cdots x_n(\mathbf{t})^{\nu_n} = \sum_{|\nu|=s} a_\nu \mathbf{t}^{\nu_1 \frac{m_1}{p_1} + \cdots + \nu_n \frac{m_n}{p_n}} =$$

$$= \sum_{|\nu|=s} b_\nu \mathbf{t}^{(\nu_1+\cdots+\nu_n)\frac{m_0}{p_0}} = \left(\sum_{|\nu|=s} b_\nu\right) \mathbf{t}^{s\frac{m_0}{p_0}} = b\mathbf{t}^{s\frac{m_0}{p_0}} = c\mathbf{t}^{1+\frac{m}{p}}.$$

where $a_\nu, b_\nu, b, c \in \mathbb{K}\{\mathbf{t}^*\}$. Hence $u = c\mathbf{t}$, which is impossible because u is a unit, and so we are done.

b) follows immediately from a).

c) Every $\varphi \in \Phi$ is defined by

$$\varphi(f) = f(x_1(\mathbf{t}), \ldots, x_n(\mathbf{t})), \quad f \in \mathcal{O}_n,$$

for unique $x_1(\mathbf{t}), \ldots, x_n(\mathbf{t})$. Thus

$$\ker(\varphi) = \{f \in \mathcal{O}_n \mid f(x_1(\mathbf{t}), \ldots, x_n(\mathbf{t})) = 0\},$$

1 Zero Sets and Zero Ideals 67

and $\ker(\varphi) \supset I$ if and only if $(x_1(\mathbf{t}), \ldots, x_n(\mathbf{t})) \in \mathcal{Z}(I)$. We hence get $c)$.

$d)$ Suppose that $f \in \ker(\varphi)$ whenever $I \subset \ker(\varphi)$, and let $x(\mathbf{t}) \in \mathcal{Z}(I)$. Then by $b)$, there is a homomorphism $\varphi \in \Phi(I)$ defined by

$$x(\mathbf{t}) = (\varphi(\mathbf{x}_1), \ldots, \varphi(\mathbf{x}_n)),$$

and consequently

$$f(x(\mathbf{t})) = f(\varphi(\mathbf{x}_1), \ldots, \varphi(\mathbf{x}_n)) = \varphi(f) = 0,$$

and so $f \in \mathcal{J}(\mathcal{Z}(I))$.

Conversely, suppose that $f \notin \ker(\varphi)$ for some φ whose kernel contains I. Then $x(\mathbf{t}) = (\varphi(\mathbf{x}_1), \ldots, \varphi(\mathbf{x}_n)) \in \mathcal{Z}(I)$, but $f(x(\mathbf{t})) = \varphi(f) \neq 0$. Thus $f \notin \mathcal{J}(\mathcal{Z}(I))$. □

Remarks and Examples 1.7 Zero sets and zero ideals are the first concern of local analytic geometry. All the machinery developed so far can be understood in geometric terms using them. We will have more occasions to do this, but here are some examples. Let $I \neq 0$ be an ideal of $\mathcal{O}_n$ and $Y = \mathcal{Z}(I) \subset D$ its zero set.

$a)$ A point $x(\mathbf{t}) \in D$ corresponds to a homomorphism $\phi \in \Phi(I)$ (Proposition 1.6), which is trivial if and only if the point is the origin $x(\mathbf{t}) = (0, \ldots, 0)$, and we exclude this case in the sequel. Hence the kernel $\mathfrak{p}$ of the homomorphism is a prime ideal different from the maximal one. Then the analytic ring $\mathcal{O}_n/\mathfrak{p}$ has dimension 1 (since ϕ induces an integral extension $\mathcal{O}_n/\mathfrak{p} \to \mathbb{K}\{\mathbf{t}^*\}$). In the converse, the question is whether every prime ideal $\mathfrak{p} \supset I$ of height $n-1$ corresponds to a point of Y. The answer depends on the coefficient field $\mathbb{K}$.

Suppose, first, $\mathbb{K} = \mathbb{C}$. In this case the analytic ring $\mathcal{O}_n/\mathfrak{p}$ has dimension 1, and by Proposition III.3.2 its normalization is $\mathbb{C}\{\mathbf{t}\}$. This gives a homomorphism $\mathcal{O}_n \to \mathcal{O}_n/\mathfrak{p} \to \mathbb{C}\{\mathbf{t}\} \to \mathbb{C}\{\mathbf{t}^*\}$, whose kernel is $\mathfrak{p}$, and hence the point of Y we sought. Thus, over $\mathbb{C}$, we can look at the zero set of I as the set of all prime ideals of height $n - 1$ containing I.

Let now $\mathbb{K} = \mathbb{R}$. Then the normalization of $\mathcal{O}_n/\mathfrak{p}$ is either $\mathbb{R}\{\mathbf{t}\}$ or $\mathbb{C}\{\mathbf{t}\}$, and it is only in the first case that we get a point of Y. We notice that the normalization is $\mathbb{R}\{\mathbf{t}\}$ if and only if $\mathcal{O}_n/\mathfrak{p}$ is a 2-real domain, in which case we call $\mathfrak{p}$ *real*. Hence, when working over $\mathbb{R}$, the zero set Y of I is seen as the set of all real prime ideals of height $n - 1$ containing I. One may ask where the non-real prime ideals have gone. The solution is easy: they are in the zero set $\widetilde{Y}$ of the ideal $\widetilde{I} = I\mathbb{C}\{\mathbf{x}\}$. This is the geometric way of defining complexification.

$b)$ We now explain the important fact that *the singular locus of Y is again a zero set*.

A *regular point of dimension 1* of Y is a point $x(\mathbf{t}) \in Y$ such that the local ring $(\mathcal{O}_n)_\mathfrak{p}/I(\mathcal{O}_n)_\mathfrak{p}$ is regular, where $\mathfrak{p}$ is the height $n - 1$ prime ideal corresponding to $x(\mathbf{t})$; otherwise $x(\mathbf{t}) \in Y$ is called a *singular point of dimension 1*. By the Regularity Jacobian Criterion (Proposition II.4.3), the set of singular points of dimension 1 of Y is the zero set of the ideal $R_1(I) + \cdots + R_n(I)$. □

$c)$ An ideal $I \subset \mathcal{O}_n$ has an *isolated singularity* if for every prime ideal $\mathfrak{p} \neq \mathfrak{m}_n$, the local ring $(\mathcal{O}_n)_\mathfrak{p}/I(\mathcal{O}_n)_\mathfrak{p}$ is regular. In the common usage the term singularity excludes the possibility of $\mathcal{O}_n/I\mathcal{O}_n$ being regular.

In case $\mathbb{K} = \mathbb{C}$ this definition means exactly that the zero set of I has no singular point of dimension 1. Indeed, this latter condition is equivalent to the fact that $(\mathcal{O}_n)_\mathfrak{p}/I(\mathcal{O}_n)_\mathfrak{p}$ is regular for every prime ideal $\mathfrak{p} \supset I$ with $\operatorname{ht}(\mathfrak{p}) = n - 1$. Choose any other prime ideal $\mathfrak{q} \supset I$, which will be contained in some $\mathfrak{p}$ with $\operatorname{ht}(\mathfrak{p}) = 1$ (this follows from Corollary II.2.5). Now, by the Regularity Jacobian Criterion (Proposition II.4.3), $\mathfrak{p} \not\supset R_s(I)$ with suitable s, and thus $\mathfrak{q} \not\supset R_s(I)$. Again by the Regularity Jacobian Criterion, $(\mathcal{O}_n)_\mathfrak{q}/I(\mathcal{O}_n)_\mathfrak{q}$ is regular of dimension $\operatorname{ht}(\mathfrak{q}) - s$.

If $\mathbb{K} = \mathbb{R}$ the preceding argument does not work, because when we choose $\mathfrak{p} \supset \mathfrak{q}$, we need $\mathfrak{p}$ to be real. For an explicit counterexample, consider the ideal $I \subset \mathbb{R}\{x, y, z\}$ generated by $f = x^2 + (y^2 + z^2)^2$. One easily checks that the zero set of I reduces to the origin, but I has not an isolated singularity: the localization at the prime ideal $\mathfrak{p} \subset \mathbb{R}\{x, y, z\}$ generated by $x, y^2 + z^2$ is not regular. Again this is reflected in the complexification: the singular point of dimension 1 that corresponds to $\mathfrak{p}$ is $(0, t, \sqrt{-1}t)$. We leave to the reader the easy exercise of stating the general result corresponding to this remark.

2 Rückert's Complex Nullstellensatz

The main result of this section is:

Proposition 2.1 *(Rückert's Nullstellensatz) Let I be an ideal of $\mathbb{C}\{\mathbf{x}\}$, where as usual $\mathbf{x} = (x_1, \ldots, x_n)$. The following assertions are equivalent:*

a) $f \in \mathcal{J}(\mathcal{Z}(I))$.

b) There is an integer $p \geq 1$ such that $f^p \in I$.

Proof. First of all, note that we can equivalently state the result in the form:
$$\mathcal{J}(\mathcal{Z}(I)) = \sqrt{I}.$$

Secondly, it is immediate that
$$\mathcal{J}(\mathcal{Z}(I)) = \mathcal{J}\left(\mathcal{Z}\left(\sqrt{I}\right)\right).$$

Now, we can write $\sqrt{I}$ as an intersection of prime ideals, namely
$$\sqrt{I} = \mathfrak{p}_1 \cap \cdots \cap \mathfrak{p}_r,$$
and by Propositions 1.3 *b)* and 1.5 *b)*
$$\mathcal{J}(\mathcal{Z}(I)) = \mathcal{J}\left(\mathcal{Z}\left(\sqrt{I}\right)\right) = \mathcal{J}(\mathcal{Z}(\mathfrak{p}_1)) \cap \cdots \cap \mathcal{J}(\mathcal{Z}(\mathfrak{p}_r)).$$

Hence, it suffices to show that
$$\mathcal{J}(\mathcal{Z}(\mathfrak{p})) = \mathfrak{p}$$

for every prime ideal $\mathfrak{p}$ of $\mathbb{C}\{\mathbf{x}\}$. Finally, by Proposition 1.6 b), this is equivalent to

$$\bigcap_{\varphi \in \Phi(\mathfrak{p})} \ker(\varphi) = \mathfrak{p}$$

for every prime ideal $\mathfrak{p}$ of $\mathbb{C}\{\mathbf{x}\}$. More explicitely, we have to show that for every $f \notin \mathfrak{p}$ there is a homomorphism of $\mathbb{C}$-algebras $\varphi : \mathbb{C}\{\mathbf{x}\} \to \mathbb{C}\{\mathbf{t}^*\}$ such that $\varphi(f) \neq 0$ and $\ker(\varphi) \supset \mathfrak{p}$.

We will deduce the last assertion from Rückert's Parametrization Theorem (Proposition II.3.4 and Remark II.3.7). After a linear change of coordinates we are under the conditions stated there, from which we pay special attention to:

a) *The canonical homomorphism* $\mathbb{C}\{\mathbf{x}'\} = A \to B = \mathbb{C}\{\mathbf{x}\}/\mathfrak{p}$ *is finite and injective, where* $\mathbf{x}' = (\mathbf{x}_1, \ldots, \mathbf{x}_d)$ *and* $d = n - \mathrm{ht}(\mathfrak{p})$.

b) *There is an irreducible polynomial* $P \in \mathbb{C}\{\mathbf{x}'\}[\mathbf{x}_{d+1}]$ *whose discriminant* $\delta \in \mathbb{C}\{\mathbf{x}'\}$ *has the property that the canonical homomorphism*

$$(\mathbb{C}\{\mathbf{x}'\}[\mathbf{x}_{d+1}]/P)_\delta \to B_\delta$$

is an isomorphism.

Since $A \to B$ is finite, the element $a = f \mod \mathfrak{p}$ is integral over A, and there is an equation

$$a^m + b_1 a^{m-1} + \cdots + b_m = 0,$$

of minimal degree m. This minimality implies that $b_m \neq 0$, because otherwise we could divide by a to get another equation of smaller degree. Hence $aa' = b \neq 0$, where

$$a' = a^{m-1} + b_1 a^{m-2} + \cdots + b_{m-1} \in B, \ b = b_m \in A.$$

Given that $\delta b = \sum a_\nu \mathbf{x}'^\nu \in \mathbb{C}\{\mathbf{x}'\}$ is not zero, it has finite order, say $p \geq 0$, and we can choose a tuple $c = (c_1, \ldots, c_d) \in \mathbb{C}^d$ such that $\sum_{|\nu|=p} a_\nu c^\nu \neq 0$. We then define a homomorphism of $\mathbb{C}$-algebras $\psi : A \to \mathbb{C}\{\mathbf{t}^*\}$ by the substitution $\mathbf{x}_i = c_i \mathbf{t}$, $1 \leq i \leq d$. By the choice of c, $\psi(\delta b) \neq 0$.

Let now $P = \mathbf{x}_{d+1}^p + a_1 \mathbf{x}_{d+1}^{p-1} + \cdots + a_p \in A[\mathbf{x}_{d+1}]$. By Newton-Puiseux's Theorem (Proposition III.4.4), the polynomial

$$P^\psi = \mathbf{x}_{d+1}^p + \psi(a_1)\mathbf{x}_{d+1}^{p-1} + \cdots + \psi(a_p) \in \mathbb{C}\{\mathbf{t}^*\}[\mathbf{x}_{d+1}]$$

has some root $x_{d+1}(\mathbf{t})$ in the quotient field $\mathbb{C}(\{\mathbf{t}^*\})$ of $\mathbb{C}\{\mathbf{t}^*\}$, and, P^ψ being monic, Proposition III.4.2 c) guarantees that $x_{d+1}(\mathbf{t}) \in \mathbb{C}\{\mathbf{t}^*\}$.

Now, since $\psi(\delta) \neq 0$, we can extend ψ to

$$\phi : (\mathbb{C}\{\mathbf{x}'\}[\mathbf{x}_{d+1}]/P)_\delta \equiv B_\delta \to \mathbb{C}(\{\mathbf{t}^*\})$$

by $\phi(x_{d+1} \bmod P) = x_{d+1}(t)$. Since B is integral over A, $\phi(B)$ is integral over $\psi(A) \subset \mathbb{C}\{t^*\}$, and the latter ring being integrally closed in its quotient field (Proposition III.4.2 c)), we conclude that $\phi(B) \subset \mathbb{C}\{t^*\}$. Hence we indeed have a homomorphism of $\mathbb{C}$-algebras $\phi : B \to \mathbb{C}\{t^*\}$.

Finally, as $\phi(aa') = \psi(b) \neq 0$, $\phi(f \bmod \mathfrak{p}) = \phi(a) \neq 0$, and the homomorphism $\varphi \in \Phi(\mathfrak{p})$ we sought is the composite of ϕ and the canonical homomorphism $\mathbb{C}\{x\} \to \mathbb{C}\{x\}/\mathfrak{p}$. This completes the proof. □

Corollary 2.2 *Let* $x = (x_1, \ldots, x_n)$, $y = (y_1, \ldots, y_p)$, $I \subset \mathbb{C}\{x\}$ *be an ideal and* $\varphi : \mathbb{C}\{y\} \to \mathbb{C}\{x\}/I$ *an analytic homomorphism. We choose series* $h_i \in \mathbb{C}\{x\}$ *such that* $\varphi(y_i) = h_i \bmod I$, $1 \leq i \leq p$. *The following assertions are equivalent:*

a) φ *is finite.*

b) $\mathcal{Z}((h_1, \ldots, h_p) + I) = \{0\}$.

Proof. Indeed, by Rückert's Nullstellensatz, *b)* is equivalent to

$$\mathfrak{m}_B = \sqrt{\varphi(\mathfrak{m}_A)B},$$

where $A = \mathbb{C}\{y\}$ and $B = \mathbb{C}\{x\}$, and this is equivalent to *a)* by the finiteness criterion of Proposition II.1.7. □

Remarks 2.3 *a)* Condition *b)* in Corollary 2.2 means that the mapping

$$\mathcal{Z}(I) \to F^p : x(t) \mapsto (h_1(x(t)), \ldots, h_p(x(t)))$$

has trivial fiber over $0 \in F^p$. This is a fundamental geometric characterization of finite maps in complex local analytic geometry.

By Proposition 1.3 *c)* we can express condition *b)* as

$$\mathcal{Z}(I) \cap \{x(t) \in F^p \mid h_1(x(t)) = \ldots = h_p(x(t)) = 0\} = \{0\}.$$

□

b) Another consequence of Rückert's Nullstellensatz is a characterization of the dimension typical of the complex case. A *hyperplane* is the zero set H of an ideal $\mathfrak{h} \subset \mathbb{C}\{x\}$ generated by homogeneous linear forms. The height d of such an ideal is clearly the minimal number of independent linear forms among the generators, and after a linear change $\mathfrak{h} = (x_1, \ldots, x_d)$; quite naturally, d is the *codimension* of H. With this terminology, from Remark II.2.7 *a)* we immediately deduce:

The dimension of an analytic ring $\mathbb{C}\{x\}/I$ *is the smallest codimension of a hyperplane* H *such that* $H \cap \mathcal{Z}(I) = \{0\}$.

Corollary 2.4 *Set* $\mathbb{K} = \mathbb{R}$ *or* $\mathbb{C}$. *The regularity ideals of a power series* $f \in \mathbb{K}\{x_1, \ldots, x_n\}$ *with* $f(0) = 0$ *are*

$$R_1(f) = \sqrt{\left(\frac{\partial f}{\partial x_1}, \ldots, \frac{\partial f}{\partial x_n}\right)}$$

and $R_s(f) = \sqrt{(f)}$ *for* $s > 1$.

Proof. It is clear from the definitions (II.4.1) that $R_s(f) = \sqrt{(f)}$ for $s > 1$ and $R_1(f) = \sqrt{(f, \partial f/\partial x_1, \ldots, \partial f/\partial x_n)}$. Hence, it suffices to prove that

$$f \in \sqrt{\left(\frac{\partial f}{\partial x_1}, \ldots, \frac{\partial f}{\partial x_n}\right)}.$$

Since the real case of this latter assertion follows from the complex one by complexification (II.5), we can assume $\mathbb{K} = \mathbb{C}$. By Rückert's Nullstellensatz, we must show that $f(x(t)) = 0$ for any $x(t) = (x_1(t), \ldots, x_n(t))$ such that $x(0) = 0$ and

$$\frac{\partial f}{\partial x_1}(x(t)) = 0, \ldots, \frac{\partial f}{\partial x_n}(x(t)) = 0.$$

These $x_i(t)$ need not have integral exponents, but after a substitution $t = t^p$ we may assume that in fact $x_i(t) \in \mathbb{C}\{t\}$, and by the chain rule

$$\frac{\partial}{\partial t} f(x(t)) = \sum_{i=1}^n \frac{\partial f}{\partial x_i}(x_i(t)) \frac{\partial x_i}{\partial t}(t) = 0,$$

and consequently

$$f(x(t)) = f(x(0)) = f(0) = 0.$$

□

Remarks 2.5 Set $\mathbb{K} = \mathbb{R}$ or $\mathbb{C}$.

a) Following Remarks and Examples 1.7 *c)*, a series $f \in \mathbb{K}\{x\}$ has a singularity if the analytic ring $\mathbb{K}\{x\}/(f)$ is not regular; this singularity is called a *hypersurface singularity*. By Proposition II.4.3 and Corollary 2.4 this happens if and only if no partial derivative $\partial f/\partial x_1, \ldots, \partial f/\partial x_n$ is a unit, that is, if and only if

$$\frac{\partial f}{\partial x_1}(0) = \cdots = \frac{\partial f}{\partial x_n}(0) = 0.$$

b) A hypersurface singularity corresponding to a series $f \in \mathbb{K}\{x\}$ is called *isolated* when $R_1(f)$ is the maximal ideal $\mathfrak{m}$ of $\mathbb{K}\{x\}$. By Proposition II.4.3, this means that for any prime ideal $\mathfrak{p} \neq \mathfrak{m}$ the local ring $\mathbb{K}\{x\}_\mathfrak{p}/f$ is regular of dimension $\operatorname{ht}(\mathfrak{p}) - 1$, which agrees with our prior definition in Remarks and Examples 1.7 *c)*. On the other hand, by the last corollary,

$$\mathfrak{m} = R_1(f) = \sqrt{\left(\frac{\partial f}{\partial x_1}, \ldots, \frac{\partial f}{\partial x_n}\right)},$$

and so the homomorphism

$$\mathbb{K} \to \mathbb{K}\{x\} \bigg/ \left(\frac{\partial f}{\partial x_1}, \ldots, \frac{\partial f}{\partial x_n}\right)$$

is finite (Corollary 2.2). This means exactly that

$$\dim_{\mathbb{K}} \left(\mathbb{K}\{x\} \bigg/ \left(\frac{\partial f}{\partial x_1}, \ldots, \frac{\partial f}{\partial x_n}\right)\right) < +\infty.$$

This dimension is called the *Milnor number* of f.

In the complex case, by Rückert's Nullstellensatz, a hypersurface singularity corresponding to a series $f \in \mathbb{C}\{x\}$ is isolated if and only if

$$\mathcal{Z}\left(f, \frac{\partial f}{\partial x_1}, \ldots, \frac{\partial f}{\partial x_n}\right) = \{0\}.$$

c) In dimension 2, square-free singularities are isolated.

Let $f \in \mathbb{K}\{x_1, x_2\}$ be square-free. We have to see that $R_1(f) \supset \mathfrak{m}$. Since in $\mathbb{K}\{x_1, x_2\}$ any prime ideal different from $\mathfrak{m}$ is principal and generated by an irreducible series, if it were $R_1(f) \not\supset \mathfrak{m}$, some irreducible power series $h \in \mathbb{K}\{x_1, x_2\}$ would divide f and all its derivatives. Thus $f = gh$, and h would divide every derivative

$$\frac{\partial f}{\partial x_i} = g\frac{\partial h}{\partial x_i} + h\frac{\partial g}{\partial x_i}.$$

Now, f being square-free, h would not divide g and, consequently, h would divide all its derivatives, which is impossible. □

d) An isolated hypersurface singularity corresponding to a series $f \in \mathbb{K}\{x\}$ is called *Morse* when its Milnor number is 1, or in other words, when the partial derivatives $\partial f/\partial x_1, \ldots, \partial f/\partial x_n$ generate the maximal ideal $\mathfrak{m}$ of $\mathbb{K}\{x\}$. By Corollary II.4.5 this is equivalent to the more familiar condition that the *Hessian matrix*

$$\left(\frac{\partial^2 f}{\partial x_i \partial x_j}(0)\right)_{1 \leq i,j \leq n}$$

has the maximum possible rank n.

3 The Homomorphism Theorem

The real Nullstellensatz is more difficult than the complex one. In this section we will obtain the key result needed to deduce it. To that end we will use the theory of formally real fields ([L XI]), which was already quoted in the proof of Newton-Puiseux's Theorem. We will use the following terminology:

Definition 3.1 *Let B be an integral domain and L its quotient field.*

a) An ordering *of B is the restriction of an ordering of L.*

b) We say that B is formally real *if there exists some ordering of B.*

Note that an ordering of L is completely determined by its restriction to B, and that B is formally real if and only if L is formally real.

Examples 3.2 *a)* The ring $\mathbb{R}\{t\}$ is formally real. In fact, it can be given exactly two different orderings, characterized by the sign of the indeterminate t.

In fact, any $f \in \mathbb{R}\{t\}$ can be written as $f = ut^p$, where u is a unit. By II.4.7 u is either a square or the opposite of one: $f = \pm v^2 t^p$, and the sign of f is completely determined by the sign of t. □

Henceforth we will always suppose $t > 0$, and the ordering is explicitely described as follows: $f = at^p + \cdots$, $a \neq 0$, is positive if and only if $a > 0$.

b) The ring $\mathbb{R}\{t^*\}$ of Puiseux series has a unique ordering in which a series $f = at^{m/p} + \cdots$, $a \neq 0$, is positive if and only if $a > 0$.

Any Puiseux series can be written in the form $f = \pm v^2 \left(t^{\frac{m}{2p}}\right)^2$, which determines completely its sign. □

Of course, we already knew this, since the field of Puiseux series with coefficients over $\mathbb{R}$ is real closed (Newton-Puiseux's Theorem). As a matter of fact, it is the real closure of $\mathbb{R}(\{t\})$ with respect to the ordering $t > 0$ described in *a)*, since it is algebraic over $\mathbb{R}(\{t\})$ by Proposition III.4.2 *d)*.

Example 3.2 *a)* is a particular case of the following:

Lemma 3.3 *Let R be a local regular ring, $\mathfrak{m}$ its maximal ideal, $k = R/\mathfrak{m}$ its residue field, and $R \to k : a \mapsto \overline{a}$ the canonical homomorphism. Let $\succ$ be an ordering in k. Then there are orderings $>$ in R such that for any unit $u \in R$: $u > 0$ if and only if $\overline{u} \succ 0$.*

Proof. By induction on the dimension d of R. Suppose first $d = 1$. Then there is an element t that generates $\mathfrak{m}$, and any $a \in R$ can be written in the form $a = ut^p$, where u is a unit and $p \geq 0$ is the largest integer with $a \in \mathfrak{m}^p$. We then set $a > 0$ if and only if $\overline{u} \succ 0$, and this is a well defined ordering of R that verifies the requirements of the statement.

The only property which is not immediate is that $a > 0$, $b > 0$ implies $a + b > 0$. To see it, write

$$a = ut^p, \quad b = vt^q, \quad u, v \notin \mathfrak{m}, \quad \overline{u} \succ 0, \overline{v} \succ 0$$

with, say, $p \leq q$. Then $a + b = wt^p$ with $w = u + vt^{q-p}$, and w is a unit with $\overline{w} \succ 0$: if $q > p$, then $\overline{w} = \overline{u} \succ 0$; if $q = p$, then $\overline{w} = \overline{u} + \overline{v} \succ 0$, since $\overline{u} \succ 0$, $\overline{v} \succ 0$.

Now let $d > 1$, and pick d elements $x_1, \ldots, x_d$ which generate $\mathfrak{m}$. The ring $R' = R/(x_1)$ is then a regular local ring of dimension $d - 1$ with residue field again k. By induction we have an ordering of R' verifying the lemma. This ordering is the restriction of an ordering $>'$ of the quotient field k_1 of R'. But k_1 is the residue field of the localization $R_1 = R_{(x_1)}$, which is a local regular ring of dimension 1. We thus know that there is some ordering $>$ of R_1 verifying the statement with $>'$ instead of $\succ$. Finally, one easily checks that the restriction of $>$ to R solves the problem. □

After this preparation we can obtain:

Proposition 3.4 *(Homomorphism Theorem) Let $\mathfrak{p}$ be a prime ideal of $\mathbb{R}\{\mathbf{x}\}$ of height $r \geq 0$, where $\mathbf{x} = (\mathbf{x}_1, \ldots, \mathbf{x}_n)$, and let $f_1, \ldots, f_p \in \mathbb{R}\{\mathbf{x}\}$. The following assertions are equivalent:*

a) *The classes $b_1 = f_1 \bmod \mathfrak{p}, \ldots, b_p = f_p \bmod \mathfrak{p}$ are positive in some ordering of the domain $B = \mathbb{R}\{\mathbf{x}\}/\mathfrak{p}$.*

b) *There is a homomorphism of $\mathbb{R}$-algebras $\phi : B \to \mathbb{R}\{\mathbf{t}^*\}$ such that $\phi(b_1) > 0, \ldots, \phi(b_p) > 0$ and $\phi(\delta \bmod \mathfrak{p}) \neq 0$ for some $\delta \in R_r(\mathfrak{p})$.*

c) *There is $x(\mathbf{t}) \in \mathcal{Z}(\mathfrak{p}) \setminus \mathcal{Z}(R_r(\mathfrak{p}))$ such that $f_1(x(\mathbf{t})) > 0, \ldots, f_p(x(\mathbf{t})) > 0$.*

(Here we set $R_0(0) = A$.)

Proof. The equivalence between *b)* and *c)* follows from Proposition 1.6 *b)* and *c)*, since the homomorphisms of $\mathbb{R}$-algebras $B \to \mathbb{R}\{t^*\}$ can be identified with the homomorphisms of $\Phi(\mathfrak{p})$.

b) $\Rightarrow$ *a)* Suppose we are given $\phi : B \to \mathbb{R}\{\mathbf{t}^*\}$ as stated in *b)*. We denote by φ the corresponding homomorphism $\mathbb{R}\{\mathbf{x}\} \to B \to \mathbb{R}\{\mathbf{t}^*\}$, and by $\mathfrak{q} \supset \mathfrak{p}$ the kernel of φ. Then $\varphi(\delta) \neq 0$ and as $\delta \in R_r(\mathfrak{p})$, we have $\mathfrak{q} \not\supset J_r(\mathfrak{p})$. By the Regularity Jacobian Criterion (Proposition II.4.3) the local ring $R = \mathbb{R}\{\mathbf{x}\}_\mathfrak{q}/\mathfrak{p}\mathbb{R}\{\mathbf{x}\}_\mathfrak{q}$ is regular. Clearly $R = B_{\ker(\phi)} \supset B$, and, since $\phi(b_i) \neq 0$, the element $b_i = f_i \bmod \mathfrak{p}$ is a unit of R.

On the other hand, the residue field k of R is the quotient field of $B/\ker(\phi)$, which embeds into $\mathbb{R}(\{\mathbf{t}^*\})$ via ϕ. The unique ordering of $\mathbb{R}(\{\mathbf{t}^*\})$ restricts, thus, to an ordering $\succ$ of k, in which the classes $b_i \bmod \ker(\phi)$ are positive, since $\phi(b_i) > 0$. Consequently, by Lemma 3.3, there is an ordering $>$ in R in which the elements b_i are positive. The restriction of $>$ to B is an ordering of B in which the b_i's are positive, and *a)* is proved.

a) $\Rightarrow$ *b)* We first note that $R_r(\mathfrak{p}) \not\subset \mathfrak{p}$: this is trivial for $r = 0$ and it follows from Lemma II.4.2 for $r > 0$. Hence, we can choose $\delta \in R_r(\mathfrak{p}) \setminus \mathfrak{p}$, so that the element $b = \delta^2 \bmod \mathfrak{p} \in B$ is positive in any ordering of B, and adding it to the b_i's, any homomorphism $\phi : B \to \mathbb{R}\{\mathbf{t}^*\}$ such that $\phi(b) > 0$ has the property that $\phi(\delta \bmod \mathfrak{p}) \neq 0$. This means that for the proof of this implication we need not care about δ.

3 The Homomorphism Theorem

We will now argue by induction on the dimension $d = n - r$ of B. If $d = 0$, then $B = \mathbb{R}$ and the assertion is trivial. We next assume $d > 0$ and the result proved for dimensions $< d$. Let L be the quotient field of B, $>$ an ordering in which the b_i's are positive and R the real closure of L with respect to $>$. We distinguish two cases:

Case 1: B is regular. If so, we may suppose $B = \mathbb{R}\{\mathbf{x}\}$ with $\mathbf{x} = (\mathbf{x}_1, \ldots, \mathbf{x}_d)$ (Lemma II.1.9). After a linear change of coordinates and by Weierstrass's Preparation Theorem (Proposition I.3.3) we have

$$b_i = u_i P_i, \quad u_i(0) \neq 0,$$

and $P_i \in \mathbb{R}\{\mathbf{x}_1, \ldots, \mathbf{x}_{d-1}\}[\mathbf{x}_d]$ is a distinguished polynomial. We remark:

If $u \in B$ is a unit and $u(0) > 0$ (resp. < 0) then u is positive (resp. negative) in any ordering of B and $\phi(u) > 0$ (resp. < 0) for any homomorphism $\phi : B \to \mathbb{R}\{\mathbf{t}^*\}$.

If $u(0) > 0$ then $u = v^2$ for some $v \in B$ (II.4.7), and consequently u is positive in any ordering; also, $\phi(u) = \phi(v)^2$ is positive for any ϕ. If $u(0) < 0$ then $u = -v^2$ and the conclusion is similar.

We thus have to find ϕ such that $\phi(P_i)$ has the same sign that P_i for all i.

We will use the following notations: $\mathbf{x}' = (\mathbf{x}_1, \ldots, \mathbf{x}_{d-1})$, $A = \mathbb{R}\{\mathbf{x}'\}$ and K for the quotient field of A. Note that $A \subset K \subset L \subset R$. Now let $\mathbf{z}$ be a new indeterminate, and consider $P_i(\mathbf{x}', \mathbf{z}) \in A[\mathbf{z}] \subset R[\mathbf{z}]$. As R is real closed we have a factorization of the form

$$P_i(\mathbf{x}', \mathbf{z}) = \prod_{j=1}^{s(i)} (\mathbf{z} - \xi_{ij}) \prod_{j=1}^{r(i)} \left((\mathbf{z} - \alpha_{ij})^2 + \beta_{ij}^2\right), \quad \beta_{ij} \neq 0, \; \xi_{ij}, \alpha_{ij}, \beta_{ij} \in R,$$

and the roots of $P_i(\mathbf{x}', \mathbf{z})$ are

$$\xi_{ij} \in R, \quad \alpha_{ij} \pm \sqrt{-1}\beta_{ij} \in R[\sqrt{-1}].$$

Since $P_i(\mathbf{x}', \mathbf{z}) \in A[\mathbf{z}]$ is monic, all these roots are integral over A, and it follows that also $\alpha_{ij}, \beta_{ij} \in R$ are integral over A. We consider the domain

$$B' = A[\xi_{ij}, \alpha_{ij}, \beta_{ij}] \subset R.$$

By the preceding remarks, B' is a finite module over A and $\sqrt{-1} \notin B'$. From Lemma III.1.1 we deduce that B' is an analytic domain over $\mathbb{R}$. Moreover, since the ξ_{ij}'s are roots of the distinguished polynomials $P_i \in A[\mathbf{x}_d]$, they belong to the maximal ideal of B' (see the beginning of the proof of Lemma III.1.1).

We now consider the elements $\xi_{ij}, \mathbf{x}_d \in R$ and choose an element $\zeta \in B'$ such that

$$\zeta - \xi_{ij} > 0 \quad \text{if and only if} \quad \mathbf{x}_d - \xi_{ij} > 0.$$

More precisely, we take

$$\zeta = \min_{ij}\{\xi_{ij}\} - \mathbf{x}_1^2, \quad \text{or} \quad \frac{1}{2}(\xi_{ij} + \xi_{i'j'}), \quad \text{or} \quad \max_{ij}\{\xi_{ij}\} + \mathbf{x}_1^2,$$

according to whether $\mathbf{x}_d$ is smaller than all the ξ_{ij}'s, or $\mathbf{x}_d$ is in between ξ_{ij} and $\xi_{i'j'}$, or $\mathbf{x}_d$ is larger than all the ξ_{ij}'s. This way ζ belongs to the maximal ideal of B', and by Lemma II.3.1, its irreducible polynomial over K is a distinguished polynomial

$$P(\mathbf{x}', \mathbf{z}) \in A[\mathbf{z}].$$

Consider the analytic ring over $\mathbb{R}$, $B^* = \mathbb{R}\{\mathbf{x}', \mathbf{z}\}/P$. By Rückert's Division Theorem (Proposition I.3.2), and since P is a distinguished polynomial, the canonical homomorphism $A[\mathbf{z}]/P \to B^*$ is an isomorphism. In particular, the embedding

$$A \to A[\mathbf{z}]/P \equiv B^*$$

is finite, and $\dim(B^*) = d - 1$. We furthermore have

$$B = \mathbb{R}\{\mathbf{x}', \mathbf{x}_d\} \to B^* = \mathbb{R}\{\mathbf{x}', \mathbf{z}\}/P \equiv A[\mathbf{z}]/P \subset K[\mathbf{z}]/P \equiv K[\zeta] \subset R,$$

where

$$\mathbf{x}_d \mapsto \mathbf{z} \bmod P \mapsto \zeta.$$

Consequently

$$P_i(\mathbf{x}', \mathbf{x}_d) \mapsto P_i^* = P_i(\mathbf{x}', \mathbf{z}) \bmod P \mapsto$$

$$\mapsto P_i(\mathbf{x}', \zeta) = \prod_{j=1}^{s(i)}(\zeta - \xi_{ij}) \prod_{j=1}^{r(i)} ((\zeta - \alpha_{ij})^2 + \beta_{ij}^2) \in R$$

and by the choice of ζ, $P_i(\mathbf{x}', \zeta)$ has the same sign as P_i. In other words, the ordering of R restricts to an ordering of B^* in which P_i^* has the same sign as P_i. Applying now the induction hypothesis, we find $\phi^* : B^* \to \mathbb{R}\{\mathbf{t}^*\}$ such that the sign of $\phi^*(P_i^*)$ is the same as the sign of P_i^*. The composite of this ϕ^* with $B \to B^*$ is the homomorphism $\phi : B \to \mathbb{R}\{\mathbf{t}^*\}$ we sought.

Finally, we note that in the argument above we tacitly assumed that some P_i had at least one real root $(s(i) > 0)$. But if this were not the case, the proof would be even simpler, because then any ζ in the maximal ideal of B' would lead to the conclusion.

Case 2: B arbitrary. Consider the ring

$$B' = B[\sqrt{b_1}, \ldots, \sqrt{b_p}] \subset R.$$

By Lemma III.1.1, B' is an analytic ring over $\mathbb{R}$, and it is enough to find $\phi' : B' \to \mathbb{R}\{\mathbf{t}^*\}$ such that $\phi(b) \neq 0$ for $b = b_1 \cdots b_p$. For, if ϕ is the restriction of ϕ' to B we get

$$\phi(b_i) = \phi'(\sqrt{b_i})^2 \geq 0 \quad \text{and} \quad \phi(b_i) \neq 0.$$

As $\dim(B') = \dim(B)$ we are reduced to the same question as in the complex case (see the proof of Proposition 2.1), namely, whether

3 The Homomorphism Theorem

for every $b \in B$, $b \neq 0$, *there is a homomorphism of* $\mathbb{R}$-*algebras* $\phi : B \to \mathbb{R}\{\mathbf{t}^*\}$ *such that* $\phi(b) \neq 0$.

Furthermore the argument given there can be repeated here, with the following modification. After a linear change the canonical homomorphism $A = \mathbb{R}\{\mathbf{x}'\} \to B$, $\mathbf{x}' = (\mathbf{x}_1, \ldots, \mathbf{x}_d)$, is finite, and we have the irreducible polynomial of $\mathbf{x}_{d+1}$ mod $\mathfrak{p}$, $P = \mathbf{x}_{d+1}^p + a_1 \mathbf{x}_{d+1}^{p-1} + \cdots + a_p \in A[\mathbf{x}_{d+1}]$, its discriminant $\delta \in A$ and another element $b \in A$. The key point is then to find a homomorphism $\phi : A \to \mathbb{R}\{\mathbf{t}^*\}$ such that $\phi(\delta b) \neq 0$ and the polynomial

$$P^\phi = \mathbf{x}_{d+1}^p + \phi(a_1)\mathbf{x}_{d+1}^{p-1} + \cdots + \phi(a_p) \in \mathbb{C}\{\mathbf{t}^*\}[\mathbf{x}_{d+1}]$$

has some root $x_{d+1}(t) \in \mathbb{R}(\{\mathbf{t}^*\})$. This was immediate in the complex case, because the field $\mathbb{C}(\{\mathbf{t}^*\})$ is algebraically closed, while $\mathbb{R}(\{\mathbf{t}^*\})$ is not. To solve this additional difficulty we use *Sturm's Theorem* ([L XI.2]):

Let $\Sigma = \{P_0, \ldots, P_s\}$ be the standard Sturm sequence *of* P, *that is:*

$$\begin{cases} P_0 = P, \ P_1 = \partial P / \partial \mathbf{x}_{d+1}, \\ P_{\ell-2} = Q_{\ell-1} P_{\ell-1} - P_\ell, \ P_\ell \neq 0, \ \deg(P_\ell) < \deg(P_{\ell-1}), \ 1 < \ell < s, \\ P_{s-1} = Q_s P_s, \end{cases}$$

where all the polynomials are in $L[\mathbf{x}_{d+1}]$. *Let*

$$M = 1 + p + a_1^2 + \cdots + a_p^2 \in A.$$

Then, since $\delta \neq 0$, P *has no multiple roots, and the number of them in the real closed field* R *equals the difference of the sing changes of the two following sequences of elements of* R:

$$\{P_0(-M), \ldots, P_s(-M)\} \quad \text{and} \quad \{P_0(M), \ldots, P_s(M)\}.$$

We now write the non-zero coefficients of the polynomials $P_\ell, Q_\ell \in L[\mathbf{x}_{d+1}]$ in the form c_i/d_i with $c_i, d_i \in A$. After this preparation we apply *Case 1* to find a homomorphism of $\mathbb{R}$-algebras $\phi : A \to \mathbb{R}\{\mathbf{t}^*\}$ such that the signs in $>$ of the elements

$$\delta, a, c_i, d_i, P_\ell(-M), P_\ell(M)$$

coincide respectively with the signs of the elements

$$\phi(\delta), \phi(a), \phi(c_i), \phi(d_i), \phi(P_\ell(-M)), \phi(P_\ell(M)).$$

We have:

a) $\Sigma^\phi = \{P_0^\phi, \ldots, P_s^\phi\}$ *is the standard Sturm sequence of* P^ϕ.

The equations that define the standard Sturm sequence are still valid after applying ϕ, because ϕ is a homomorphism that does not map to zero any denominator d_i. Moreover, the conditions on the degrees are also preserved, since ϕ does not map to zero any numerator c_i, including those corresponding to the terms of higher degree of the P_ℓ's. We similarly see that

b) The discriminant of P^ϕ is $\phi(\delta) \neq 0$.

c) $\phi(M) = 1 + p + \phi(a_1)^2 + \cdots + \phi(a_p)^2$.

d) The sign changes in the sequences

$$\{P_0(-M), \ldots, P_s(-M)\} \quad \text{and} \quad \{P_0^\phi(-\phi(M)), \ldots, P_s^\phi(-\phi(M))\}$$

$$(\text{resp. } \{P_0(M), \ldots, P_s(M)\} \quad \text{and} \quad \{P_0^\phi(\phi(M)), \ldots, P_s^\phi(\phi(M))\})$$

coincide.

After these remarks we apply Sturm's Theorem to the polynomial P^ϕ and deduce that the number of roots of P^ϕ in the real closed field $\mathbb{R}(\{t^*\})$ equals the number of real roots of P in R. But P has at least the root $x_{d+1} \mod \mathfrak{p} \in B \subset R$, and consequently P^ϕ has at least one root $x_{d+1}(\mathbf{t}) \in \mathbb{R}(\{\mathbf{t}^*\})$ as wanted.

As was explained before, once this x_{d+1} is available, the proof follows to the end as in the complex case. □

Example 3.5 The regularity ideal is essential for the equivalence in the preceding result. For instance, let $\mathfrak{p} \subset \mathbb{R}\{\mathbf{x}, \mathbf{y}, \mathbf{z}\}$ be the ideal generated by $\mathbf{x}^2 - \mathbf{z}\mathbf{y}^2$ and consider the series $f = -\mathbf{z}$. Then $b = f \mod \mathfrak{p} = -\mathbf{x}^2 \mod \mathfrak{p}/\mathbf{y}^2 \mod \mathfrak{p}$ is negative in all orderings, but for $x(\mathbf{t}) = (0, 0, -\mathbf{t}) \in \mathcal{Z}(\mathfrak{p})$ we have $f(x(\mathbf{t})) = t > 0$. Of course the reason is that $R_1(\mathfrak{p}) = (\mathbf{x}, \mathbf{y})$, and so $x(\mathbf{t}) \in \mathcal{Z}(R_1(\mathfrak{p}))$. This example is the famous *Whitney's Umbrella*.

4 Risler's Real Nullstellensatz

The real counterpart of Proposition 2.1 is:

Proposition 4.1 *(Risler's Nullstellensatz) Let I be an ideal of $\mathbb{R}\{\mathbf{x}\}$, where as usual $\mathbf{x} = (\mathbf{x}_1, \ldots, \mathbf{x}_n)$. The following assertions are equivalent:*

a) $f \in \mathcal{J}(\mathcal{Z}(I))$.

b) *There are an integer $p \geq 1$ and power series $g_1, \ldots, g_s \in \mathbb{R}\{\mathbf{x}\}$ such that*

$$f^{2p} + g_1^2 + \cdots + g_s^2 \in I$$

Once we have obtained the Homomorphism Theorem (Proposition 3.4) the only missing ingredient for the proof of the real Nullstellensatz is a notion that substitutes the radical of an ideal and formalizes condition *b)* in the above statement.

4 Risler's Real Nullstellensatz

Proposition and Definition 4.2 *Let A be a commutative ring with unit, and $I \subset A$ an ideal. The set of all elements $f \in A$ such that $f^{2p} + g_1^2 + \cdots + g_s^2 \in I$ for some $p \geq 1$ and some $g_1, \ldots, g_s \in A$ is an ideal of A, called the* real radical *of I and denoted by $\sqrt[r]{I}$.*

Proof. The tricky part is that if $f, g \in \sqrt[r]{I}$, then $f + g \in \sqrt[r]{I}$. To see this, write

$$f^{2p} + a \in I, \quad g^{2q} + b \in I,$$

where a, b are sums of squares of A. Now, if $p \geq q$, we have:

$$\left((f+g)^2 + (f-g)^2\right)^{2p} = \left(2f^2 + 2g^2\right)^{2p} = \alpha f^{2p} + \beta g^{2q},$$

and

$$\left((f+g)^2 + (f-g)^2\right)^{2p} = (f+g)^{4p} + \gamma,$$

where α, β, γ are also sums of squares of A. It follows

$$(f+g)^{4p} + \gamma + \alpha a + \beta b = \alpha(f^{2p} + a) + \beta(f^{2q} + b) \in I,$$

and so $f + g \in \sqrt[r]{I}$. □

We finally come to the

Proof of Proposition 4.1: According to the definition of the real radical, the statement can be equivalently formulated as

$$\mathcal{J}(\mathcal{Z}(I)) = \sqrt[r]{I}.$$

To start with, we note that

$$\mathcal{J}(\mathcal{Z}(I)) = \mathcal{J}\left(\mathcal{Z}\left(\sqrt[r]{I}\right)\right).$$

Indeed, if $f \in \sqrt[r]{I}$ we have an expression $f^{2p} + g_1^2 + \cdots + g_s^2 \in I$. Then, if $x(\mathbf{t}) \in \mathcal{Z}(I)$, we get

$$f(x(\mathbf{t}))^{2p} + g_1(x(\mathbf{t}))^2 + \cdots + g_s(x(\mathbf{t}))^2 = 0.$$

This is a sum of squares of elements of $\mathbb{R}(\{\mathbf{t}^*\})$, which is a formally real field, and consequently, all the summands must be zero, so that $f(x(\mathbf{t})) = 0$. This means that $\mathcal{Z}(I) \subset \mathcal{Z}\left(\sqrt[r]{I}\right)$, and consequently

$$\mathcal{J}(\mathcal{Z}(I)) \supset \mathcal{J}\left(\mathcal{Z}\left(\sqrt[r]{I}\right)\right).$$

The other inclusion follows from the general properties of $\mathcal{Z}$ and $\mathcal{J}$, since $I \subset \sqrt[r]{I}$.

Now the ideal $\sqrt[r]{I}$ is clearly radical and has a decomposition

$$\sqrt[r]{I} = \mathfrak{p}_1 \cap \cdots \cap \mathfrak{p}_r$$

into prime ideals. Then
$$\mathcal{J}(\mathcal{Z}(I)) = \mathcal{J}(\mathcal{Z}(\mathfrak{p}_1)) \cap \cdots \cap \mathcal{J}(\mathcal{Z}(\mathfrak{p}_r)),$$
and we have to see that $\mathcal{J}(\mathcal{Z}(\mathfrak{p}_i)) = \mathfrak{p}_i$ for all i.

Fix $i = 1, \ldots, r$ and put $B = \mathbb{R}\{\mathbf{x}\}/\mathfrak{p}_i$. We claim that B is a formally real domain. We recall *Artin-Schreier's Criterion* ([L XI.2]):

> B is formally real if and only if any equation $z_1^2 + \cdots + z_s^2 = 0$ has only trivial solutions in B.

To check this criterion in our case, let $h_1, \ldots, h_s \in \mathbb{R}\{\mathbf{x}\}$ be such that
$$h_1^2 + \cdots + h_s^2 = 0 \bmod \mathfrak{p}_i.$$
We choose an element $h \in \bigcap_{j \neq i} \mathfrak{p}_j \setminus \mathfrak{p}_i$ and obtain
$$(hh_1)^2 + \cdots + (hh_s)^2 \in \bigcap_{j \neq i} \mathfrak{p}_j \cap \mathfrak{p}_i = \sqrt[r]{I}.$$
Hence from the very definition we get
$$hh_1, \ldots, hh_s \in \sqrt[r]{I} \subset \mathfrak{p}_i,$$
and since $h \notin \mathfrak{p}_i$, we conclude $h_1, \ldots, h_s \in \mathfrak{p}_i$. In other words
$$h_1 = \cdots = h_s = 0 \bmod \mathfrak{p}_i,$$
as required.

We finally show that $\mathcal{J}(\mathcal{Z}(\mathfrak{p}_i)) \subset \mathfrak{p}_i$ (the other inclusion is obvious). Fix an ordering $>$ in B and let $f \notin \mathfrak{p}_i$. Then $f^2 \bmod \mathfrak{p}_i$ is positive in $>$, and by the Homomorphism Theorem there is $x(\mathbf{t}) \in \mathcal{Z}(\mathfrak{p}_i)$ such that $f(x(\mathbf{t}))^2 > 0$. Then $f(x(\mathbf{t})) \neq 0$ and $f \notin \mathcal{J}(\mathcal{Z}(\mathfrak{p}_i))$. □

In this way we have characterized the zero ideals. In particular for prime ideals we obtain:

Corollary 4.3 *Let $\mathfrak{p}$ be a prime ideal of height $r > 0$ of $R\{\mathbf{x}\}$, $\mathbf{x} = (\mathbf{x}_1, \ldots, \mathbf{x}_n)$. Then the following assertions are equivalent:*

a) $\mathcal{J}(\mathcal{Z}(\mathfrak{p})) = \mathfrak{p}$.

b) The domain $B = \mathbb{R}\{\mathbf{x}\}/\mathfrak{p}$ is formally real.

c) There is a homomorphism of $\mathbb{R}$-algebras $\phi : B \to \mathbb{R}\{\mathbf{t}^*\}$ such that $\phi(\delta \bmod \mathfrak{p}) \neq 0$ for some $\delta \in R_r(\mathfrak{p})$.

d) $\mathcal{Z}(\mathfrak{p}) \setminus \mathcal{Z}(R_r(\mathfrak{p})) \neq \emptyset$.

Proof. By Risler's Nullstellensatz (Proposition 4.1) *a)* is equivalent to the following:

If $f^{2p} + g_1^2 + \cdots + g_s^2 \in \mathfrak{p}$ then $f \in \mathfrak{p}$.

This, in turn, can be reformulated as:

If $f^{2p} + g_1^2 + \cdots + g_s^2 = 0 \bmod \mathfrak{p}$, then $f \in \mathfrak{p}$.

But $\mathfrak{p}$ is prime, and so $f \in \mathfrak{p}$ if and only if $f^p \in \mathfrak{p}$. Hence the last condition is exactly Artin-Schreier's Criterion. Consequently, *a)* is equivalent to *b)*.

On the other hand, conditions *b)*, *c)* and *d)* are equivalent by the homomorphism theorem (Proposition 3.2) for $f_i = 1$. □

Note that *a)* is always true in the complex case, while in the real one we have the algebraic conditions *b)* and *c)*. Condition *b)* is a strengthening of the notion of a 2-real domain which was introduced in connection with complexifications (Proposition and Definition II.5.6). In general it is strictly stronger: $\mathbb{R}\{\mathbf{x},\mathbf{y},\mathbf{z}\}/(\mathbf{x}^2 + \mathbf{y}^2 + \mathbf{z}^2)$ is a 2-real domain, but it is not formally real. However in dimension 1 we have:

Proposition 4.4 *(Risler) Let B be an analytic ring over $\mathbb{R}$, which is a domain of dimension 1. Then B is 2-real if and only if it is formally real.*

Proof. The "if" part is always true, so suppose that B is 2-real. Then its normalization is isomorphic to $\mathbb{K}\{\mathbf{t}\}$ (Proposition III.3.2) with $\mathbb{K} = \mathbb{R}$ or $\mathbb{C}$. But since B is 2-real, $\sqrt{-1} \notin \mathbb{K}\{\mathbf{t}\}$, and so $\mathbb{K} = \mathbb{R}$. Thus, the quotient field of B is isomorphic to that of $\mathbb{R}\{\mathbf{t}\}$, and as the latter ring is a formally real domain, we are done. □

5 Hilbert's 17th Problem

Let $W = \mathbb{R}\{\mathbf{t}^*\}$ be the ring of convergent Puiseux series, $F = \mathbb{R}(\{\mathbf{t}^*\})$ its quotient field and $U = \mathfrak{m}^*$ its maximal ideal. Fix $n \geq 1$ and consider the set $D = U \times \cdots \times U \subset F^n$ (1.1)

Definition 5.1 *Let Y be a subset of D. A power series $f \in \mathbb{R}\{\mathbf{x}\}$, $\mathbf{x} = (\mathbf{x}_1, \ldots, \mathbf{x}_n)$, is called* positive semidefinite on Y *if $f(x(\mathbf{t})) \geq 0$ for every $x(\mathbf{t}) \in Y$.*

Now Hilbert's 17th Problem makes sense for power series with real coefficients, and its solution is, as expected, that a positive semidefinite power series is always a sum of squares. But this has to be formulated in a suitable way.

Proposition 5.2 *Let I be an ideal of $\mathbb{R}\{\mathbf{x}\}$, $\mathbf{x} = (\mathbf{x}_1, \ldots, \mathbf{x}_n)$, and $f \in \mathbb{R}\{\mathbf{x}\}$. Then the following assertions are equivalent:*

a) f is positive semidefinite on $\mathcal{Z}(I)$.

b) There are $p \geq 1$ and $h_1, \ldots, h_r, g_1, \ldots, g_s \in \mathbb{R}\{\mathbf{x}\}$ such that

$$f\left(f^{2p} + h_1^2 + \cdots + h_r^2\right) = g_1^2 + \cdots + g_s^2 \bmod I.$$

c) There are $h, g_1, \ldots, g_s \in \mathbb{R}\{x\}$ such that
$$fh^2 = g_1^2 + \cdots + g_s^2 \mod I$$
and $\mathcal{Z}(h) \subset \mathcal{Z}(f)$.

Proof. $a) \Rightarrow b)$ If $f(0) \neq 0$, then $a)$ implies $f(0) > 0$, and f is a square in $\mathbb{R}\{x\}$. Thus we can assume $f(0) = 0$. Consider a new indeterminate z and the canonical inclusion $\mathbb{R}\{x\} \subset \mathbb{R}\{x, z\}$. Let $f_1, \ldots, f_m \in \mathbb{R}\{x\}$ be generators of I. We claim that
$$z \in \mathcal{J}\left(\mathcal{Z}\left(f + z^2, f_1, \ldots, f_m\right)\right).$$
Indeed, if $(x(t), z(t)) \in \mathcal{Z}\left(z^2 + f, f_1, \ldots, f_m\right)$, then $f_i(x(t)) = 0$ for all i, so that $x(t) \in \mathcal{Z}(I)$ and by $a)$ $f(x(t)) \geq 0$. Moreover $z(t)^2 + f(x(t)) = 0$ and consequently $z(t) = 0$. Thus our claim is proved. Now, by Risler's Nullstellensatz (Proposition 4.1), there are $q \geq 1$ and $\alpha, \beta_j, F_i \in \mathbb{R}\{x, z\}$ such that
$$z^{2q} + \sum_i F_i(x, z)^2 = \alpha(x, z)(z^2 + f(x)) + \sum_j \beta_j(x, z) f_j(x).$$

Furthermore, multiplying by z^2 if necessary, we may assume that q is odd: $q = 2p+1$ with $p \geq 1$.

We next note that every series $F \in \mathbb{R}\{x, z\}$ can be uniquely written in the form
$$F(x, z) = G(x, z^2) + z H(x, z^2);$$
we call G the 0-component of F and H the 1-component. In our case we get
$$z^{2q} + \sum_i \left(G_i(x, z^2) + z H_i(x, z^2)\right)^2 =$$
$$\left(\alpha_0(x, z^2) + z\alpha_1(x, z^2)\right)(z^2 + f(x)) + \sum_j \left(\beta_{j0}(x, z^2) + z\beta_{j1}(x, z^2)\right) f_j(x)$$
and comparing the 0-components of both sides of this equality it follows
$$z^{2q} + \sum_i \left(G_i(x, z^2)^2 + z^2 H_i(x, z^2)^2\right) = \alpha_0(x, z^2)(z^2 + f(x)) + \sum_j \beta_{j0}(x, z^2) f_j(x).$$

Recalling that $q = 2p+1$, we rewrite this equation in the form:
$$z^2 \left(z^{4p} + \sum_i H_i(x, z^2)^2\right) + \sum_i G_i(x, z^2)^2 =$$
$$\alpha_0(x, z^2)(z^2 + f(x)) + \sum_j \beta_{j0}(x, z^2) f_j(x).$$

On the other hand the homomorphism
$$\mathbb{R}\{x, z\} \to \mathbb{R}\{x, z\} : T(x, z) \mapsto T(x, z^2)$$

5 Hilbert's 17th Problem

is injective. For, it is finite (1 and z generate the second ring as a module over the first), and, since both rings are domains of the same dimension, its kernel must be zero. Hence, from the last equation we deduce

$$z\left(z^{2p} + \sum_i H_i(\mathbf{x}, z)^2\right) + \sum_i G_i(\mathbf{x}, z)^2 = \alpha_0(\mathbf{x}, z)(z + f(\mathbf{x})) + \sum_j \beta_{j0}(\mathbf{x}, z) f_j(\mathbf{x}).$$

Finally, since $f(0) = 0$, we can substitute $z = -f(\mathbf{x})$ to obtain the equation of the statement.

$b) \Rightarrow c)$ It is enough to take $h = f^{2p} + h_1^2 + \cdots + h_r$, which obviously verifies the condition required:

$$fh^2 = (f^{2p} + h_1^2 + \cdots + h_r)(g_1^2 + \cdots + g_s^2) \bmod I,$$

where the right hand side is again a sum of squares.

$c) \Rightarrow a)$ Suppose $fh^2 = g_1^2 + \cdots + g_s^2 \bmod I$ with $\mathcal{Z}(h) \subset \mathcal{Z}(f)$, and let $x(t) \in \mathcal{Z}(I)$. If $f(x(t)) = 0$ there is nothing to prove. If $f(x(t)) \neq 0$, we also have, by the condition on h, $h(x(t)) \neq 0$, and since

$$f(x(t)) h(x(t))^2 = g_1(x(t))^2 + \cdots + g_s(x(t))^2 \geq 0,$$

we conclude $f(x(t)) \geq 0$. □

Remarks and Examples 5.3 *a)* The preceding result shows that positive semidefinite functions are sums of squares, allowing denominators of course. Nevertheless, these denominators are not arbitrary.

b) Conversely a sum of squares with arbitrary denominator need not be positive semidefinite. For instance, let $I = \mathfrak{p} \subset \mathbb{R}\{\mathbf{x}, \mathbf{y}, \mathbf{z}\}$ be the ideal generated by $\mathbf{x}^2 - \mathbf{z}\mathbf{y}^2$ and consider the series $f = z$. Then $fy^2 = x^2 \bmod \mathfrak{p}$, but for $x(t) = (0, 0, -t) \in \mathcal{Z}(\mathfrak{p})$ we have $f(x(t)) = -t < 0$. We are to see that this fact is true in general, and the obstruction to positiveness lies in the regularity ideals.

After the preceding remarks we will characterize the sum of squares with arbitrary denominators. For the sake of simplicity we restrict ourselves to the case of prime ideals, and leave to the reader the exercise of formulating the corresponding result for arbitrary ideals.

Proposition 5.4 *Let $\mathfrak{p}$ be a prime ideal of height $r \geq 0$ of $\mathbb{R}\{\mathbf{x}\}$, $\mathbf{x} = (\mathbf{x}_1, \ldots, \mathbf{x}_n)$. The following assertions are equivalent:*

a) f is positive semidefinite on $\mathcal{Z}(\mathfrak{p}) \setminus \mathcal{Z}(R_r(\mathfrak{p}))$.

b) There are $h, g_1, \ldots, g_s \in \mathbb{R}\{\mathbf{x}\}$, $h \notin \mathfrak{p}$, such that

$$fh^2 = g_1^2 + \cdots + g_s^2 \bmod \mathfrak{p}$$

Proof. Set $B = \mathbb{R}\{\mathbf{x}\}/\mathfrak{p}$ and let L be the quotient field of B.

Suppose first $\mathcal{Z}(\mathfrak{p}) \setminus \mathcal{Z}(R_r(\mathfrak{p})) = \emptyset$. Then B is not formally real (Proposition 4.3) and -1 is a sum of squares in L. On the other hand, $b = f \bmod \mathfrak{p} \in B$ can be written in the form
$$b = \frac{1}{4}\left((b+1)^2 - (b-1)^2\right),$$
and, -1 being a sum of squares in L, b is a sum of squares, too. This shows that under our initial assumption both conditions *a)* and *b)* hold, and we are done.

Let now $\mathcal{Z}(\mathfrak{p}) \setminus \mathcal{Z}(R_r(\mathfrak{p})) \neq \emptyset$. Then B is formally real (Proposition 4.3) and the element $b = f \bmod \mathfrak{p} \in B$ is a sum of squares in L if and only if b is positive in all orderings of B ([L XI.2]). By the Homomorphism Theorem (Proposition 3.4) the latter assertion is equivalent to *a)*. □

Of course all these differences disappear in the regular case:

Corollary 5.5 *(Risler) Let $f \in \mathbb{R}\{\mathbf{x}\}$, $\mathbf{x} = (\mathbf{x}_1, \ldots, \mathbf{x}_n)$. The following assertions are equivalent:*

a) f is positive semidefinite on D.

b) There are $h, g_1, \ldots, g_s \in \mathbb{R}\{\mathbf{x}\}$, $h \neq 0$, such that
$$fh^2 = g_1^2 + \cdots + g_s^2.$$

c) There are $h, g_1, \cdots, g_s \in \mathbb{R}\{\mathbf{x}\}$ such that
$$fh^2 = g_1^2 + \ldots + g_s^2$$
and $\mathcal{Z}(h) \subset \mathcal{Z}(f)$.

Remarks and Examples 5.6 *a)* In general, denominators cannot be avoided. Indeed, Motzkin found the following example. Consider the *Arithmetic-Geometric Identity*
$$(a+b+c)^3 - 27abc = \frac{27}{4}c(a-b)^2 + \frac{1}{4}(4a+4b+c)(a+b-2c)^2.$$
After the substitution $a = \mathbf{x}^4\mathbf{y}^2$, $b = \mathbf{y}^4\mathbf{z}^2$, $c = \mathbf{z}^4\mathbf{x}^2$ we get a series $h \in \mathbb{R}\{\mathbf{x},\mathbf{y},\mathbf{z}\}$ which factorizes into $h = fg$ with
$$f(\mathbf{x},\mathbf{y},\mathbf{z}) = \alpha - 3\beta, \quad g(\mathbf{x},\mathbf{y},\mathbf{z}) = \alpha^2 + \alpha\beta + \beta^2,$$
$$\alpha = \mathbf{x}^4\mathbf{y}^2 + \mathbf{y}^4\mathbf{z}^2 + \mathbf{z}^4\mathbf{x}^2, \quad \beta = \mathbf{x}^2\mathbf{y}^2\mathbf{z}^2.$$
Clearly f, g, h are sums of squares, and we claim that to express f as a sum of squares the denominators are essential.

Let $f = \sum_i f_i^2$ for some $f_i \in \mathbb{R}\{\mathbf{x},\mathbf{y},\mathbf{z}\}$. Since f is an homogeneous polynomial of degree 6, looking at the expansions of the f_i's we can suppose they are homogeneous polynomials of degree 3. Then

5 Hilbert's 17th Problem

- For $y = z = 0$ we get $0 = \sum_i (a_i x^3)^2$, and so $a_i = 0$ and x^3 does not appear in any f_i. Similarly, y^3, z^3 do not appear either.

- For $z = 0$ we get $x^4 y^2 = \sum_i (a_i x^2 y + b_i xy^2)^2$. Hence $b_i = 0$ and xy^2 does not appear in any f_i. Analogously, neither $x^2 z$ nor yz^2 appear in any f_i.

Summarizing,
$$f_i = a_i x^2 y + b_i y^2 z + c_i z^2 x + d_i xyz,$$
and consequently, $\sum_i d_i^2 = -3$, which is impossible. □

b) Denominators are not necessary in dimension ≤ 2. For dimension 1 the proof is easy: any $f \in \mathbb{R}\{t\}$ can be written as $f = \varepsilon u^2 t^q$ with $\varepsilon = \pm 1$ and $u(0) \neq 0$. If f is positive semidefinite, we have

- $0 \leq f(+t) = \varepsilon u^2 t^q$, and so $\varepsilon = +1$.
- $0 \leq f(-t) = u(-t)^2(-t)^q$, and so q is even, say $q = 2p$.

Hence $f = (ut^p)^2$ is a square without any denominator. □

We end this section with the proof of the fact stated above that denominators are not needed in dimension 2.

Proposition 5.7 *(Bochnak-Risler) Every power series $f \in \mathbb{R}\{x_1, x_2\}$ which is positive semidefinite is a sum of two squares in $\mathbb{R}\{x_1, x_2\}$.*

Proof. We first write $f = x_1^p g$ with $g(0, x_2) \neq 0$. Then g is regular of some order with respect to x_2 and by Weierstrass Preparation Theorem there is a distinguished polynomial $P \in \mathbb{R}\{x_1\}[x_2]$ and a unit $u \in \mathbb{R}\{x_1, x_2\}$ such that $f = ux_1^p P$.

We now factorize P into irreducible components according to Remark II.2.2, and, after associating factors of even multiplicity, we can write $P = P_1 P_2$, where P_1 is a square in $\mathbb{R}\{x_1, x_2\}$ and P_2 has not multiple factors. We claim that P_2 has no root in the field of Puiseux series $F = \mathbb{R}(\{t^*\})$.

Suppose otherwise, and let ξ the biggest of such roots. Since P_2 is a distinguished polynomial, $\xi \in \mathbb{R}\{t^*\}$, and furthermore ξ belongs to the maximal ideal $U = \mathfrak{m}^*$ of $\mathbb{R}\{t^*\}$. By the choice of ξ there are elements ζ and ζ' in U such that:

a) $\zeta' < \xi < \zeta$,

b) ζ and ζ' are bigger than all the other roots of P_2, and

c) ζ and ζ' are different from all the roots of P.

We can now factorize P_2 in $F[x_2]$ as:
$$P_2 = (x_2 - \xi) \prod_i (x_2 - \xi_i) \prod_j \left((x_2 - \alpha_j)^2 + \beta_j^2\right), \quad \beta_j \neq 0,$$
and it follows easily that

$$f(\mathbf{t},\zeta)f(\mathbf{t},\zeta') < 0,$$

which is impossible because f is positive semidefinite. Thus the claim is proved.

From the claim we deduce that P_2 is positive semidefinite, and since P_1 is a square, P is positive semidefinite, too. Then for any $x(\mathbf{t}) = (\pm\mathbf{t}, \theta)$, $\theta \in U$, with $P(x(\mathbf{t})) \neq 0$ we get

$$f(x(\mathbf{t})) = (\pm\mathbf{t})^p u(x(\mathbf{t})) P(x(\mathbf{t})) > 0,$$

which implies $(\pm 1)^p u(0,0) > 0$ and consequently $u(0,0) > 0$ and p is even. In other words, we have

$$f = g^2 Q_1 \cdots Q_s,$$

where $g \in \mathbb{R}\{\mathbf{x}_1, \mathbf{x}_2\}$ and every $Q_i \in \mathbb{R}\{\mathbf{x}_1\}[\mathbf{x}_2]$ is an irreducible distinguished polynomial without roots in F.

Next, the identity

$$(a^2 + b^2)(c^2 + d^2) = (ac - bd)^2 + (ad + bc)^2$$

shows that it is enough to prove that every $Q = Q_i$ is a sum of 2 squares in $\mathbb{R}\{\mathbf{x}_1, \mathbf{x}_2\}$. To see this, consider the analytic ring $A = \mathbb{R}\{\mathbf{x}_1, \mathbf{x}_2\}/Q$. If this ring where 2-real, its normalization would be isomorphic to $\mathbb{R}\{\mathbf{t}\}$, and we would get a homomorphism $\mathbb{R}\{\mathbf{x}_1, \mathbf{x}_2\} \to \mathbb{R}\{\mathbf{t}\}$ such that $\mathbf{x}_1 \mapsto \mathbf{t}^q$, $\mathbf{x}_2 \mapsto x_2(\mathbf{t})$ and $Q(\mathbf{t}^q, x_2(\mathbf{t})) = 0$. Then $\xi = x_2(\mathbf{t}^{1/q})$ would be a root of Q, which has none. Hence, A is not 2-real, and so its complexification is not a domain. In other words, Q is reducible in $\mathbb{C}\{\mathbf{x}_1, \mathbf{x}_2\}$. We can thus factorize Q in $\mathbb{C}\{\mathbf{x}_1, \mathbf{x}_2\}$, and by the properties of conjugation (II.5.2), one immediately gets:

$$Q = H\overline{H}, \quad H = h_1 + \sqrt{-1} h_2, \quad h_1, h_2 \in \mathbb{R}\{\mathbf{x}_1, \mathbf{x}_2\}.$$

Consequently

$$Q = h_1^2 + h_2^2,$$

which ends the proof. $\square$

V Approximation Theory

Summary. The central result of this chapter is M. Artin's Approximation Theorem of formal solutions of analytic systems, which we obtain in Section 3. To prove it we need a generalization of the classical Implicit Functions Theorem due to Tougeron (Section 1). This generalization has its own independent interest, which we illustrate in Section 2 with several consequences concerning the equivalence of power series and polynomials. Next we deduce (Section 4) the excellent behaviour of analytic rings under completion. Finally, we introduce in Section 5 the Nash rings, which are the smallest subrings of analytic rings that share with them all the nice properties proved so far.

1 Tougeron's Implicit Functions Theorem

Set $\mathbf{x} = (\mathbf{x}_1, \ldots, \mathbf{x}_n)$, $\mathbf{y} = (\mathbf{y}_1, \ldots, \mathbf{y}_p)$, and

$$A = \mathbb{K}\{\mathbf{x}\} \text{ (resp. } \mathbb{K}[[\mathbf{x}]]), \quad B = \mathbb{K}\{\mathbf{x}, \mathbf{y}\} \text{ (resp. } \mathbb{K}[[\mathbf{x}, \mathbf{y}]]).$$

We fix an element $F = (F_1, \ldots, F_q) \in B^q$ with $F(0,0) = 0$, and look for a solution $y_1 = \mathbf{y}_1(\mathbf{x}), \ldots, y_p = \mathbf{y}_p(\mathbf{x})$ of the system

$$F(\mathbf{x}, \mathbf{y}_1, \ldots, \mathbf{y}_p) = 0,$$

under assumptions milder than the ones of the classical Implicit Functions Theorem. Of course those conditions will involve the Jacobian matrix

$$\lambda = \left(\frac{\partial F_i}{\partial \mathbf{y}_j}(\mathbf{x}, 0)\right).$$

(1.1) The matrix λ defines a homomorphism of A-modules $\lambda : A^p \to A^q$. Namely, let $\{\varepsilon_1, \ldots, \varepsilon_p\}$ and $\{e_1, \ldots, e_q\}$ be the canonical bases of A^p and A^q. Then

$$\lambda(\varepsilon_j) = \sum_{i=1}^{q} \frac{\partial F_i}{\partial \mathbf{y}_j}(\mathbf{x}, 0) e_i = \left(\frac{\partial F_1}{\partial \mathbf{y}_j}(\mathbf{x}, 0), \ldots, \frac{\partial F_q}{\partial \mathbf{y}_j}(\mathbf{x}, 0)\right).$$

Furthermore, we consider the A-module

$$M = A^q / \text{Im}(\lambda).$$

With these notations, an element $\delta \in A$ belongs to the annihilator $\text{Ann}(M)$ of M if and only if there are power series $\alpha_{ij} \in A$ such that

$$\delta e_i = \lambda \left(\sum_{j=1}^{p} \alpha_{ij}\varepsilon_j \right) = \lambda(\alpha_{i1}, \ldots, \alpha_{ip});$$

then, if $\mu : A^q \to A^p$ is the homomorphism defined by the matrix (α_{ij}), we can rewrite the latter condition in the form

$$\lambda \circ \mu = \delta \cdot \mathrm{Id}_{A^q}.$$

We now introduce new indeterminates

$$\mathbf{y}^{(i)} = (\mathbf{y}_1^{(i)}, \ldots, \mathbf{y}_p^{(i)}), \ 1 \leq i \leq r,$$

where r is any fixed positive integer. In this situation

Lemma 1.2 *Let $\delta_1, \ldots, \delta_r \in \mathrm{Ann}(M)$. Then, there are convergent power series $Y^{(i)} = Y^{(i)}(\mathbf{x}, \mathbf{y}^{(1)}, \ldots, \mathbf{y}^{(r)})$, $1 \leq i \leq r$, such that*

$$F\left(\mathbf{x}, \sum_{i=1}^{r} \delta_i Y^{(i)}\right) = F(\mathbf{x}, 0) + \lambda\left(\sum_{i=1}^{r} \delta_i \mathbf{y}^{(i)}\right),$$

$$Y^{(1)}(\mathbf{x}, 0, \ldots, 0) = \cdots = Y^{(r)}(\mathbf{x}, 0, \ldots, 0) = 0.$$

Proof. Consider further indeterminates

$$\mathbf{z} = (z_1, \ldots, z_p); \ \mathbf{z}^{(i)} = (z_1^{(i)}, \ldots, z_p^{(i)}), \ 1 \leq i \leq r.$$

Using the Taylor expansion with respect to $\mathbf{z}$ we obtain

$$F(\mathbf{x}, \mathbf{z}) = F(\mathbf{x}, 0) + \lambda(\mathbf{z}) + \sum_{\ell, m} G_{\ell, m} z_\ell z_m,$$

with $G_{\ell, m} \in B^q$. By means of the substitution $\mathbf{z} = \sum_{i=1}^{r} \delta_i \mathbf{z}^{(i)}$ we then get

$$F\left(\mathbf{x}, \sum_{i=1}^{r} \delta_i \mathbf{z}^{(i)}\right) = F(\mathbf{x}, 0) + \lambda\left(\sum_{i=1}^{r} \delta_i \mathbf{z}^{(i)}\right) + \sum_{ij} Y_{ij}(\mathbf{x}, \mathbf{z}^{(1)}, \ldots, \mathbf{z}^{(r)})\delta_i \delta_j,$$

$$Y_{ij}(\mathbf{x}, 0, \ldots, 0) = \frac{\partial Y_{ij}}{\partial z_m^{(\ell)}}(\mathbf{x}, 0, \ldots, 0) = 0,$$

with $Y_{ij} \in E^q$ and $E = \mathbb{K}\{\mathbf{x}, \mathbf{z}^{(1)}, \ldots, \mathbf{z}^{(r)}\}$.

Now, by the remark preceding the statement, there are homomorphisms of A-modules $\mu_1, \ldots, \mu_r : A^q \to A^p$ such that

$$\lambda \circ \mu_j = \delta_j \cdot \mathrm{Id}_{A^q}, \ 1 \leq j \leq r.$$

These μ_j extend naturally to homomorphisms $E^q \to E^p$, and λ to a homomorphism $E^p \to E^q$; these extensions are denoted by the same letters. We then put

1 Tougeron's Implicit Functions Theorem

$$a_i = z^{(i)} + \sum_{j=1}^{r} \mu_j(Y_{ij}), \ 1 \leq i \leq r,$$

and get

$$\lambda(a_i) = \lambda(z^{(i)}) + \sum_{j=1}^{r} \lambda \circ \mu_j(Y_{ij}) = \lambda(z^{(i)}) + \sum_{j=1}^{r} Y_{ij} \delta_j.$$

Hence

$$F\left(\mathbf{x}, \sum_{i=1}^{r} \delta_i \mathbf{z}^i\right) = F(\mathbf{x}, 0) + \lambda \left(\sum_{i=1}^{r} \delta_i a^i\right). \tag{1}$$

On the other hand, the system

$$a_i(\mathbf{x}, \mathbf{z}^{(1)}, \ldots, \mathbf{z}^{(r)}) = \mathbf{y}^{(i)}, \ 1 \leq i \leq r,$$

can be solved by means of the classical Implicit Function Theorem (Proposition II.4.6 b)). In fact, since $Y_{ij}(\mathbf{x}, 0, \ldots, 0) = 0$ it follows that

$$a_i(\mathbf{x}, 0, \ldots, 0) = 0,$$

and, since $\frac{\partial Y_{ij}}{\partial z_m^{(\ell)}}(\mathbf{x}, 0, \ldots, 0) = 0$, that

$$\frac{\partial a_i}{\partial z_m^{(\ell)}}(\mathbf{x}, 0, \ldots, 0) = 1.$$

Hence, there are convergent series $Y^{(i)} = z^{(i)}(\mathbf{x}, \mathbf{y}^{(1)}, \ldots, \mathbf{y}^{(r)})$ such that

$$a_i(\mathbf{x}, Y^{(1)}, \ldots, Y^{(r)}) = \mathbf{y}^{(i)}, \ 1 \leq i \leq r.$$

Moreover, the $Y^{(i)}(\mathbf{x}, 0, \ldots, 0)$'s are a solution of

$$a_i(\mathbf{x}, 0, \ldots, 0) = 0, \ 1 \leq i \leq r,$$

and this system has only the trivial solution (again by the classical Implicit Functions Theorem). We conclude

$$Y^{(i)}(\mathbf{x}, 0, \ldots, 0) = 0, \ 1 \leq i \leq r,$$

and, substituting $z^{(i)} = Y^{(i)}$ in (1) we obtain the formula of the statement. □

Finally we obtain:

Proposition 1.3 *(Tougeron's Implicit Functions Theorem)* With all the notations introduced above, suppose $q \leq p$ and let $I \subset A$ be the ideal generated by the q-minors of the matrix λ and $I' \subset A$ another ideal. Then, if

$$F_1(\mathbf{x}, 0), \ldots, F_q(\mathbf{x}, 0) \in I'I^2,$$

there are $y_1(\mathbf{x}), \ldots, y_p(\mathbf{x}) \in I'I$ such that

$$F(\mathbf{x}, y_1(\mathbf{x}), \ldots, y_p(\mathbf{x})) = 0.$$

Proof. Let δ be a q-minor of λ, say the one consisting of the first q rows and columns. With the notations of 1.1 we have

$$\delta = \det(\lambda(\varepsilon_1), \ldots, \lambda(\varepsilon_q)).$$

The expansion of this determinant gives

$$\delta e_i = \sum_{j=1}^{n} (-1)^{j+1} \det(\lambda(\varepsilon_1), \ldots, a_j^*, \ldots, \lambda(\varepsilon_q))\lambda(\varepsilon_j),$$

where $a_j^* = e_i$. This shows that $\delta A^q \subset \text{Im}(\lambda)$, and so, $\delta \in \text{Ann}(M)$. In particular,

$$I'I^2 \oplus \cdots \oplus I'I^2 \subset (\text{Im}(\lambda))\,I.$$

Let now $\delta_1, \ldots, \delta_r$ be the q-minors of λ, which generate I. From the hypothesis we get

$$F(\mathbf{x}, 0) = \lambda \left(\sum_{i=1}^{r} \delta_i \beta^{(i)} \right),$$

where $\beta^{(i)} = (\beta_1^{(i)}, \ldots, \beta_r^{(i)})$, $\beta_j^{(i)} \in I'$. We make $\mathbf{y}^{(i)} = -\beta^{(i)}$ in the formula of Lemma 1.2, to obtain $F(\mathbf{x}, y(\mathbf{x})) = 0$ with

$$y(\mathbf{x}) = (y_1(\mathbf{x}), \ldots, y_p(\mathbf{x})) = \sum_{i=1}^{r} \delta_i Y^{(i)}(\mathbf{x}, -\beta^{(1)}, \ldots, -\beta^{(r)}).$$

Using the Taylor expansion with respect to $\mathbf{y}^{(1)}, \ldots, \mathbf{y}^{(r)}$ of this power series, and taking into account that $Y^{(i)}(\mathbf{x}, 0, \ldots, 0) = 0$, one easily checks that all the monomials in the $y_1(\mathbf{x}), \ldots, y_p(\mathbf{x})$ are generated by the elements $\delta_i \beta_j^{(\ell)} \in I'I$. Hence

$$y_1(\mathbf{x}), \ldots, y_p(\mathbf{x}) \in \bigcap_{s \geq 0} (I'I + \mathfrak{m}^s) = I'I$$

(the last equality by Krull's Theorem). The proof is thus complete. $\square$

2 Equivalence of Power Series

Here we will obtain some basic results concerning the so-called equivalence problem. Let $\mathbb{K} = \mathbb{R}$ or $\mathbb{C}$ and consider the convergent power series ring $\mathbb{K}\{\mathbf{x}\}$, where $\mathbf{x} = (x_1, \ldots, x_n)$; the maximal ideal of $\mathbb{K}\{\mathbf{x}\}$ will be denoted by $\mathfrak{m}$. As usual we omit the discussion of the formal case, which would be analogous.

2 Equivalence of Power Series

Definition 2.1 *Two power series $f, g \in \mathbb{K}\{\mathbf{x}\}$ are called* equivalent *when there are power series $h_1, \ldots, h_n \in \mathbb{K}\{\mathbf{x}\}$ such that*

$$h_1(0) = \cdots = h_n(0) = 0, \quad \frac{D(h_1, \ldots, h_n)}{D(\mathbf{x}_1, \ldots, \mathbf{x}_n)}(0) \neq 0$$

and

$$f(h_1, \ldots, h_n) = g.$$

This is an equivalence relation by the Inverse Function Theorem (Proposition II.4.6 a)). On the other hand, the above giving of $h_1, \ldots, h_n$ is the same as the giving of an analytic isomorphism of $\mathbb{K}\{\mathbf{x}\}$ that maps f to g (Propositions II.1.3, II.1.8 and Corollary II.4.5).

Now the problem is to recognize which power series are equivalent to a given one $f \in \mathbb{K}\{\mathbf{x}\}$. To address it, we introduce a modified Jacobian ideal, namely:

$$I_f = \left(\frac{\partial f}{\partial \mathbf{x}_1}, \ldots, \frac{\partial f}{\partial \mathbf{x}_n} \right),$$

which we already used to define the Milnor number (Remark IV.2.5 b)).

We notice that $I_f \not\subset \mathfrak{m}$ if and only if some partial derivative is a unit, say $\partial f/\partial \mathbf{x}_1(0) \neq 0$. Applying then the Inverse Function Theorem to $f, \mathbf{x}_2, \ldots, \mathbf{x}_n$ we see that f is equivalent to $\mathbf{x}_1$, and one easily checks that a power series is equivalent to $\mathbf{x}_1$ if and only if its order is 1. Consequently we will always suppose $I_f \subset \mathfrak{m}$.

We next prove a useful criterion:

Lemma 2.2 *Let $f \in \mathbb{K}\{\mathbf{x}\}$ be a power series with $f(0) = 0$ and $I_f \subset \mathfrak{m}$. Then, f is equivalent to every $g \in \mathbb{K}\{\mathbf{x}\}$ such that $f - g \in \mathfrak{m} I_f^2$.*

Proof. Consider new indeterminates $\mathbf{y} = (\mathbf{y}_1, \ldots, \mathbf{y}_n)$ and the equation

$$F(\mathbf{x}, \mathbf{y}) = f(\mathbf{x} + \mathbf{y}) - g(\mathbf{x}) = 0.$$

We have the Jacobian matrix

$$\lambda = \left(\frac{\partial F}{\partial \mathbf{y}_i}(\mathbf{x}, 0) \right)_{1 \leq i \leq n} = \left(\frac{\partial f}{\partial \mathbf{x}_i}(\mathbf{x}, 0) \right)_{1 \leq i \leq n},$$

and we can apply Tougeron's Implicit Functions Theorem (Proposition 1.3) with $q = 1$, $I = I_f$ and $I' = \mathfrak{m}$: there are $y_1(\mathbf{x}), \ldots, y_n(\mathbf{x}) \in \mathfrak{m} I_f$ such that

$$F(\mathbf{x}, y_1(\mathbf{x}), \ldots, y_n(\mathbf{x})) = 0,$$

that is,

$$f(\mathbf{x}_1 + y_1(\mathbf{x}), \ldots, \mathbf{x}_n + y_n(\mathbf{x})) = g(\mathbf{x}).$$

Finally, the power series we are looking for are

$$h_i = \mathbf{x}_i + y_i(\mathbf{x}),\ 1 \le i \le n.$$

Indeed,
$$\frac{\partial h_i}{\partial \mathbf{x}_j} = \frac{\partial y_i}{\partial \mathbf{x}_j} \text{ if } j \ne i, \quad \frac{\partial h_i}{\partial \mathbf{x}_i} = 1 + \frac{\partial y_i}{\partial \mathbf{x}_i}.$$

Since $y_i \in \mathfrak{m} I_f \subset \mathfrak{m}^2$,
$$\frac{\partial y_i}{\partial \mathbf{x}_j}(0) = 0,$$

which gives
$$\frac{D(h_1, \ldots, h_n)}{D(\mathbf{x}_1, \ldots, \mathbf{x}_n)}(0) = 1.$$

□

A nice application of this lemma is:

Proposition 2.3 *Let $f \in \mathbb{K}\{\mathbf{x}\}$ have an isolated singularity with Milnor number μ, and $g \in \mathbb{K}\{\mathbf{x}\}$ any power series with $f - g \in \mathfrak{m}^{2\mu+1}$. Then f and g are equivalent.*

Proof. We recall that $\mu = \dim_{\mathbb{K}} (\mathbb{K}\{\mathbf{x}\}/I_f) \ge 1$, and so $I_f \subset \mathfrak{m}$. We also know that $\mathfrak{m} = \sqrt{I_f}$, and since power series rings are noetherian, there is some ℓ such that $\mathfrak{m}^\ell \subset I_f$. In fact, we can take $\ell = \mu$. Indeed, consider the sequence
$$\mathbb{K}\{\mathbf{x}\} \supset \mathfrak{m} + I_f \supset \cdots \supset \mathfrak{m}^\mu + I_f \supset I_f.$$

As the dimension of $\mathbb{K}\{\mathbf{x}\}/I_f$ over $\mathbb{K}$ is μ, one of the inclusions above must be an equality, and so we get
$$\mathfrak{m}^k + I_f = \mathfrak{m}^{k+1} + I_f$$

for some $k \le \mu$. Multiplying by $\mathfrak{m}$, we obtain
$$\mathfrak{m}^\mu + I_f = \mathfrak{m}^{\mu+1} + I_f = \cdots = \mathfrak{m}^\ell + I_f = I_f,$$

and consequently $\mathfrak{m}^\mu \subset I_f$.

Once this is shown,
$$f - g \in \mathfrak{m}^{2\mu+1} \subset \mathfrak{m} I_f^2,$$

and the result follows from the preceding lemma. □

Remarks 2.4 The last proposition contains two famous facts:

a) Every isolated hypersurface singularity is algebraic. More precisely, if $f \in \mathbb{K}\{\mathbf{x}\}$ has an isolated singularity with Milnor number μ, then f is equivalent to a polynomial of degree $\le 2\mu$.

This follows from Proposition 2.3 by taking g to be the sum of the terms of degree $\le 2\mu$ of the Taylor expansion of f. □

2 Equivalence of Power Series

b) (Morse's Lemma) Let $f \in \mathbb{K}\{\mathbf{x}\}$ have a Morse singularity. By Proposition 2.3, f is equivalent to the quadratic form

$$Q(\mathbf{x}) = \sum_{i,j} \frac{\partial^2 f}{\partial \mathbf{x}_i \partial \mathbf{x}_j}(0) \mathbf{x}_i \mathbf{x}_j.$$

Then, by the elementary theory of symmetric matrices, after a linear change of coordinates we obtain:

$$Q(\mathbf{x}) = \mathbf{x}_1^2 + \cdots + \mathbf{x}_n^2, \text{ if } \mathbb{K} = \mathbb{C},$$

$$Q(\mathbf{x}) = \mathbf{x}_1^2 + \cdots + \mathbf{x}_s^2 - \mathbf{x}_{s+1}^2 - \cdots - \mathbf{x}_n^2, \text{ if } \mathbb{K} = \mathbb{R}.$$

Remarks 2.5 *a)* In case the singularity is not isolated the equivalence problem becomes much more difficult, but Tougeron's Implicit Functions Theorem provides a systematic way to obtain equivalence criteria that generalize Lemma 2.2. An important fact that can be proved in this way is that *every power series $f \in \mathbb{K}\{\mathbf{x}_1, \ldots, \mathbf{x}_n\}$, $n \geq 2$, is equivalent to a polynomial in two indeterminates $g \in \mathbb{K}\{\mathbf{x}_1, \ldots, \mathbf{x}_{n-2}\}[\mathbf{x}_{n-1}, \mathbf{x}_n]$.*

Such a systematic approach goes beyond the scope of this section, and it will not be discussed here, but we can use Lemma 2.2 as it is to deduce the result stated above for square-free singularities. Indeed, let f be square-free, $f(0) = 0$ and $f \neq 0$. We consider again the ideal I_f and claim that

$$\mathrm{ht}(I_f) \geq 2.$$

This is proved using the trick of Remark IV.2.5 *c)*. Suppose $\mathrm{ht}(I_f) = 1$, so that there is a height 1 prime ideal $\mathfrak{p} \supset I_f$. Since the ring $\mathbb{K}\{\mathbf{x}\}$ is factorial (Proposition II.2.1), $\mathfrak{p}$ is principal, say generated by an irreducible power series h. Thus

$$\sqrt{I_f} \subset (h).$$

Furthermore $f \in \sqrt{I_f}$ by Corollary IV.2.4, and we deduce that h divides f and its derivatives $\partial f / \partial \mathbf{x}_1, \ldots, \partial f / \partial \mathbf{x}_n$. Hence $f = gh$ for some power series g and, f being square-free, h does not divide g. Thus

$$\frac{\partial f}{\partial \mathbf{x}_i} = g \frac{\partial h}{\partial \mathbf{x}_i} + h \frac{\partial g}{\partial \mathbf{x}_i}.$$

It follows that h divides all its derivatives, which is impossible. Thus the proof of the claim is finished.

We consider now the ideal

$$J = \mathfrak{m} I_f^2 \subset \mathfrak{m}.$$

Since $I_f \supset J \supset I_f^3$, we see that $\mathrm{ht}(J) = \mathrm{ht}(I_f) \geq 2$, and after a linear change of coordinates (II.2.3) we may assume that the canonical homomorphism

$$\mathbb{K}\{\mathbf{x}_1, \ldots, \mathbf{x}_{n-2}\} \to \mathbb{K}\{\mathbf{x}\}/J$$

is finite. There is then a monic polynomial
$$P(t) = P(x_1, \ldots, x_{n-2}, t) \in \mathbb{K}\{x_1, \ldots, x_{n-2}\}[t]$$
such that
$$P(x_{n-1}), P(x_n) \in J$$
(take P to be the product of two equations of integral dependence of x_{n-1}, x_n mod J). Then, by Rückert's Division Theorem,
$$f = g + QP(x_{n-1}) + Q'P(x_n),$$
where $g \in \mathbb{K}\{x\}$ is a polynomial in x_{n-1}, x_n and $Q, Q' \in \mathbb{K}\{x\}$. We thus have $f - g \in J = \mathfrak{m}I_f^2$ and the proof ends by applying Lemma 2.2. □

b) If we restrict our attention to two indeterminates, a) says that *every power series in two indeterminates is equivalent to a polynomial*, and we have given a proof of this for square-free power series. In fact, this is also a consequence of Proposition 2.3, since square-free power series in two indeterminates have only isolated singularities (Remarks and Examples IV.2.5 c)).

3 M. Artin's Approximation Theorem

As usual, we consider the coefficient field $\mathbb{K} = \mathbb{R}$ or $\mathbb{C}$, and indeterminates $x = (x_1, \ldots, x_n)$, $y = (y_1, \ldots, y_p)$. We will denote by $\mathfrak{m}$ the maximal ideal of a ring of convergent power series, and by $\hat{\mathfrak{m}}$ the maximal ideal of the corresponding ring of formal power series.

This section is devoted to the main result of the chapter:

Proposition 3.1 *(M. Artin's Approximation Theorem)* Let $f = (f_1, \ldots, f_q) \in \mathbb{K}\{x, y\}^q$ be such that $f(0, 0) = 0$. Consider a solution $\hat{y}(x) = (\hat{y}_1(x), \ldots, \hat{y}_q(x)) \in \mathbb{K}[[x]]^q$ of the system $f(x, y) = 0$. Then for every integer $\alpha \geq 1$ there exists a solution $y(x) = (y_1(x), \ldots, y_q(x)) \in \mathbb{K}\{x\}^q$ of $f(x, y) = 0$ such that
$$y(x) \equiv \hat{y}(x) \bmod \hat{\mathfrak{m}}^\alpha.$$

Proof. We argue by induction on n. For $n = 0$ there is nothing to prove, hence we suppose $n \geq 1$ and the result proved for fewer that n x_k's.

The proof will proceed now by descending induction on the height of the ideal $I \subset \mathbb{K}\{x, y\}$ generated by the f_i's. Our first step is:

A) *Reduction to prime ideals.*

We have a decomposition $\sqrt{I} = \mathfrak{p}_1 \cap \cdots \cap \mathfrak{p}_r$ for some prime ideals
$$\mathfrak{p}_k = (g_{k1}, \ldots, g_{kq_k}) \subset \mathbb{K}\{x, y\}.$$

3 M. Artin's Approximation Theorem

We consider the elements $g^{(k)} = (g_{k1}, \ldots, g_{kq_k}) \in \mathbb{K}\{\mathbf{x}, \mathbf{y}\}^{q_k}$, and claim that $\hat{y}(\mathbf{x})$ is a solution of some of the analytic systems $g^{(k)}(\mathbf{x}, \mathbf{y}) = 0$. Suppose we had $g_{k\ell_k}(\mathbf{x}, \hat{y}(\mathbf{x})) \neq 0$ for $1 \leq k \leq r$. There would then be $m \geq 1$ such that

$$h = (g_{1\ell_1} \cdots g_{r\ell_r})^m \in I,$$

that is,

$$h(\mathbf{x}, \mathbf{y}) = h_1(\mathbf{x}, \mathbf{y})f_1(\mathbf{x}, \mathbf{y}) + \cdots + h_q(\mathbf{x}, \mathbf{y})f_q(\mathbf{x}, \mathbf{y})$$

for some $h_1, \ldots, h_q$. Thus, making $\mathbf{y} = \hat{y}(\mathbf{x})$, we would get non-zero on the left hand side and zero on the right hand side. This contradiction shows that $\hat{y}(\mathbf{x})$ must be a solution of some of the systems $g^{(k)}(\mathbf{x}, \mathbf{y}) = 0$. Whence the reduction A) is done.

We now start our descending induction on $s = \text{ht}(I)$. If I has the maximal height $s = n + p$, the only prime ideal containing I is the maximal ideal, and for this everything is trivial. By the preceding reduction, the result follows for I. Let, next, $s < n + p$ and assume the result to be proved for heights $> s$. We start again with a simplification of the problem:

B) *Reduction to the case when I is generated by s elements $h_1, \ldots, h_s$, and there is a Jacobian*

$$\delta = \frac{D(h_1, \ldots, h_s)}{D(y_{k_1}, \ldots, y_{k_s})}$$

such that $\delta(\mathbf{x}, \hat{y}(\mathbf{x})) \neq 0$.

By the previous reduction we can assume that I is a prime ideal of height s which will be denoted by $\mathfrak{p}$. There are then $h_1, \ldots, h_s \in \mathfrak{p}$ and indeterminates $z_1, \ldots, z_s$ among $x_1, \ldots, x_n, y_1, \ldots, y_p$ such that the Jacobian

$$\delta' = \frac{D(h_1, \ldots, h_s)}{D(z_1, \ldots, z_s)} \notin \mathfrak{p}$$

(Lemma II.4.2). In particular, $\text{ht}(\mathfrak{p} + \delta'\mathbb{K}\{\mathbf{x}, \mathbf{y}\}) > s$ and if $\delta'(\mathbf{x}, \hat{y}(\mathbf{x})) = 0$ we can apply the induction hypothesis to conclude the proof. Hence let $\delta'(\mathbf{x}, \hat{y}(\mathbf{x})) \neq 0$. On the other hand, since h_i belongs to the ideal generated by the f_j's, we have $h_i(\mathbf{x}, \hat{y}(\mathbf{x})) = 0$ and derivating:

$$\frac{\partial h_i}{\partial x_j}(\mathbf{x}, \hat{y}(\mathbf{x})) = -\sum_{\ell=1}^{p} \frac{\partial h_i}{\partial y_\ell}(\mathbf{x}, \hat{y}(\mathbf{x})) \frac{\partial \hat{y}_\ell(\mathbf{x})}{\partial x_j}.$$

Consequently, if, say,

$$z_1 = x_1, \ldots, z_m = x_m, z_{m+1} = y_1, \ldots, z_s = y_t,$$

a straightforward computation gives

$$0 \neq \delta'(\mathbf{x}, \hat{y}(\mathbf{x})) =$$
$$= (-1)^m \sum_{\ell_1,\ldots,\ell_m} \frac{\partial \hat{y}_{\ell_1}(\mathbf{x})}{\partial \mathbf{x}_1} \cdots \frac{\partial \hat{y}_{\ell_m}(\mathbf{x})}{\partial \mathbf{x}_m} \frac{D(h_1,\ldots,h_s)}{D(\mathbf{y}_{\ell_1},\ldots,\mathbf{y}_{\ell_m},\mathbf{y}_1,\ldots,\mathbf{y}_t)}(\mathbf{x}, \hat{y}(\mathbf{x})).$$

Hence, some of the Jacobians in the right hand side must be different from zero, and we can suppose without loss of generality

$$0 \neq \delta(\mathbf{x}, \hat{y}(\mathbf{x})), \quad \delta = \frac{D(h_1,\ldots,h_s)}{D(\mathbf{y}_1,\ldots,\mathbf{y}_s)} \notin \mathfrak{p}.$$

By the Regularity Jacobian Criterion (Proposition II.4.3) the elements $h_1,\ldots,h_s$ generate the maximal ideal $\mathfrak{p}\mathbb{K}\{\mathbf{x},\mathbf{y}\}_\mathfrak{p}$ of the localization $\mathbb{K}\{\mathbf{x},\mathbf{y}\}_\mathfrak{p}$. Thus, there exists $h \notin \mathfrak{p}$ such that $h\mathfrak{p} \subset (h_1,\ldots,h_s) \subset \mathbb{K}\{\mathbf{x},\mathbf{y}\}$, that is:

$$hf_i = \lambda_{i1}h_1 + \cdots + \lambda_{is}h_s \tag{1}$$

We again have $\mathrm{ht}(\mathfrak{p}+h\mathbb{K}\{\mathbf{x},\mathbf{y}\}) > s$, so that by induction hypothesis we may suppose $h(\mathbf{x}, \hat{y}(\mathbf{x})) \neq 0$.

After all this preparation, let $y(\mathbf{x}) \in \mathbb{K}\{\mathbf{x}\}^p$ be a solution of the system $h_1(\mathbf{x},\mathbf{y}) = \cdots = h_s(\mathbf{x},\mathbf{y}) = 0$ such that

$$y_i(\mathbf{x}) = \hat{y}_i(\mathbf{x}) \bmod \widehat{\mathfrak{m}}^\alpha.$$

We then deduce using Taylor expansions that

$$h(\mathbf{x}, y(\mathbf{x})) = h(\mathbf{x}, \hat{y}(\mathbf{x})) \bmod \widehat{\mathfrak{m}}^\alpha.$$

Indeed, $h(\mathbf{x}, y(\mathbf{x})) - h(\mathbf{x}, \hat{y}(\mathbf{x}))$ belongs to the ideal generated by the differences $y_i(\mathbf{x}) - \hat{y}_i(\mathbf{x})$, which belong all to $\widehat{\mathfrak{m}}^\alpha$. Once we know this, for α large enough it follows $h(\mathbf{x}, y(\mathbf{x})) \neq 0$. From this and the equation (1) above we conclude that $y(\mathbf{x})$ is a solution of the initial system $f_1(\mathbf{x},\mathbf{y}) = \cdots = f_q(\mathbf{x},\mathbf{y}) = 0$. The proof of this reduction is finished.

Henceforth we assume the conditions of reduction B), and consider the convergent power series $\Delta = \mathbf{x}_n^\alpha \delta^2$. Then:

C) *There exist convergent power series* $z_1(\mathbf{x}),\ldots,z_p(\mathbf{x}) \in \mathbb{K}\{\mathbf{x}\}$ *with*

$$z_1(\mathbf{x}) = \hat{y}_1(\mathbf{x}),\ldots,z_p(\mathbf{x}) = \hat{y}_p(\mathbf{x}) \bmod \widehat{\mathfrak{m}}^\alpha$$

and

$$h_1(\mathbf{x}, z(\mathbf{x})),\ldots,h_s(\mathbf{x}, z(\mathbf{x})) \in \Delta(\mathbf{x}, z(\mathbf{x}))\mathbb{K}\{\mathbf{x}\},$$

where $z(\mathbf{x}) = (z_1(\mathbf{x}),\ldots,z_p(\mathbf{x}))$.

3 M. Artin's Approximation Theorem

To prove this, up to a linear change of coordinates we can suppose that $\Delta(\mathbf{x}, \hat{y}(\mathbf{x}))$ is regular of order s with respect to $\mathbf{x}_n$, and consequently $\Delta(\mathbf{x}, \hat{y}(\mathbf{x}))^\alpha$ is regular of order αs with respect to $\mathbf{x}_n$. Setting $\mathbf{x}' = (\mathbf{x}_1, \ldots, \mathbf{x}_{n-1})$ and applying Rückert's Division Theorem we get

$$\hat{y}_j(\mathbf{x}) = \hat{q}_j(\mathbf{x})\Delta(\mathbf{x}, \hat{y}(\mathbf{x}))^\alpha + \hat{z}_j(\mathbf{x}),$$
$$\hat{z}_j(\mathbf{x}) = \sum_{\ell=1}^{\alpha s} \mathbf{x}_n^{\alpha s - \ell}(\hat{y}_{j\ell}(\mathbf{x}') + y_{j\ell}^0),$$
$$\hat{q}_j(\mathbf{x}) \in \mathbb{K}[[\mathbf{x}]],\ \hat{y}_{j\ell}(\mathbf{x}') \in \mathbb{K}[[\mathbf{x}']],\ \hat{y}_{j\ell}(0) = 0,\ y_{j\ell}^0 \in \mathbb{K}.$$

Furthermore

$$\hat{z}_j(0) = \hat{y}_j(0) - \hat{q}_j(0)\Delta(0, \hat{y}(0))^\alpha = 0,$$

that is,

$$y_{j,\alpha s}^0 = 0.$$

We write $\hat{z}(\mathbf{x}) = (\hat{z}_1(\mathbf{x}), \ldots, \hat{z}_p(\mathbf{x}))$ and since the order of the power series $\Delta(\mathbf{x}, \hat{y}(\mathbf{x}))^\alpha$ is $\geq \alpha$, it holds

$$\hat{y}_1(\mathbf{x}) = \hat{z}_1(\mathbf{x}), \ldots, \hat{y}_p(\mathbf{x}) = \hat{z}_p(\mathbf{x}) \bmod \widehat{\mathfrak{m}}^\alpha.$$

On the other hand, looking at the Taylor expansion of the series $\Delta(\mathbf{x}, \hat{y}(\mathbf{x})) - \Delta(\mathbf{x}, \hat{z}(\mathbf{x}))$, we see that it belongs to the ideal generated by the differences

$$\hat{y}_1(\mathbf{x}) - \hat{z}_1(\mathbf{x}), \ldots, \hat{y}_p(\mathbf{x}) - \hat{z}_p(\mathbf{x}).$$

But this ideal is contained in the one generated by $\Delta(\mathbf{x}, \hat{y}(\mathbf{x}))^\alpha$, and so

$$\Delta(\mathbf{x}, \hat{y}(\mathbf{x})) - \Delta(\mathbf{x}, \hat{z}(\mathbf{x})) = \hat{h}(\mathbf{x})\Delta(\mathbf{x}, \hat{y}(\mathbf{x}))^\alpha, \quad \hat{h}(\mathbf{x}) \in \mathbb{K}[[\mathbf{x}]],$$

or equivalently

$$\Delta(\mathbf{x}, \hat{z}(\mathbf{x})) = \hat{u}(\mathbf{x})\Delta(\mathbf{x}, \hat{y}(\mathbf{x})), \quad \hat{u}(\mathbf{x}) = 1 - \hat{h}(\mathbf{x})\Delta(\mathbf{x}, \hat{y}(\mathbf{x}))^{\alpha-1},$$

and for $\alpha \geq 2$, $\hat{u}(0) \neq 0$ and $\hat{u}$ is a unit. Thus, the power series $\Delta(\mathbf{x}, \hat{z}(\mathbf{x}))$ is regular of order s with respect to $\mathbf{x}_n$ as $\Delta(\mathbf{x}, \hat{y}(\mathbf{x}))$ is.

We now introduce new indeterminates $\mathbf{y}_{j\ell}$ and consider the polynomials

$$z_j = \sum_{\ell=1}^{\alpha s} \mathbf{x}_n^{\alpha s - \ell}(\mathbf{y}_{j\ell} - y_{j\ell}^0).$$

We put $z = (z_1, \ldots, z_p)$ and notice that $\Delta(\mathbf{x}, z) \in \mathbb{K}\{\mathbf{x}, \mathbf{y}_{j\ell}\}$ is regular of order s with respect to $\mathbf{x}_n$ as $\Delta(\mathbf{x}, \hat{z})$ is, because $z_j(\mathbf{x}, 0) = \hat{z}_j(\mathbf{x}, 0)$. Hence by Rückert's Division Theorem

$$h_i(\mathbf{x}, z) = Q_i(\mathbf{x}, \mathbf{y}_{j\ell})\Delta(\mathbf{x}, z) + \sum_{m=1}^{s} \mathbf{x}_n^{s-m} g_{im}(\mathbf{x}', \mathbf{y}_{j\ell}),$$

where $Q_i(\mathbf{x}, \mathbf{y}_{j\ell})$ and the $g_{im}(\mathbf{x}', \mathbf{y}_{j\ell})$'s are convergent power series. Making the substitution $\mathbf{y}_{j\ell} = \hat{y}_{j\ell}(\mathbf{x}')$ we obtain:

$$h_i(\mathbf{x}, \hat{z}(\mathbf{x})) = Q_i(\mathbf{x}, \hat{y}_{j\ell}(\mathbf{x}'))\Delta(\mathbf{x}, \hat{z}(\mathbf{x})) + \sum_{m=1}^{s} x_n^{s-m} g_{im}(\mathbf{x}', \hat{y}_{j\ell}(\mathbf{x}')). \qquad (2)$$

We now recall that $\hat{y}(\mathbf{x})$ is a formal solution of $f(\mathbf{x}, \mathbf{y}) = 0$, so that $h_i(\mathbf{x}, \hat{y}(\mathbf{x})) = 0$ and

$$h_i(\mathbf{x}, \hat{z}(\mathbf{x})) = h_i(\mathbf{x}, \hat{z}(\mathbf{x})) - h_i(\mathbf{x}, \hat{y}(\mathbf{x})).$$

Using again the Taylor expansion, and that the differences $\hat{z}_j(\mathbf{x}) - \hat{y}_j(\mathbf{x})$ belong to the ideal generated by $\Delta(\mathbf{x}, \hat{y}(\mathbf{x}))^\alpha$, it follows at once that $h_i(\mathbf{x}, \hat{z}(\mathbf{x}))$ also belongs to that ideal. In particular, $h_i(\mathbf{x}, \hat{z}(\mathbf{x}))$ is divisible by $\Delta(\mathbf{x}, \hat{y}(\mathbf{x}))$, and consequently also by $\Delta(\mathbf{x}, \hat{z}(\mathbf{x})) = \hat{u}(\mathbf{x})\Delta(\mathbf{x}, \hat{y}(\mathbf{x}))$, because $\hat{u}$ is a unit. This means that the remainder in the division of equation (2) must be zero, or in other words

$$g_{im}(\mathbf{x}', \hat{y}_{j\ell}(\mathbf{x}')) = 0$$

for $1 \leq m \leq s$, $1 \leq i \leq q$. We thus have a solution $\hat{y}_{j\ell}(\mathbf{x}') \in \mathbb{K}[[\mathbf{x}']]$ of an analytic system with $n-1$ x_k's, and by induction hypothesis, there is a solution $y_{j\ell}(\mathbf{x}') \in \mathbb{K}\{\mathbf{x}'\}$ such that $y_{j\ell} \equiv \hat{y}_{j\ell} \bmod \widehat{\mathfrak{m}}^\alpha$. Substituting $\mathbf{y}_{j\ell}$ by $y_{j\ell}$ in the z_j's we get

$$z_j(\mathbf{x}) = \sum_{\ell=1}^{\alpha s} x_n^{\alpha s - \ell}(y_{j\ell}(\mathbf{x}') - y_{j\ell}^0) \in \mathbb{K}\{\mathbf{x}\},$$

and

$$z_j(\mathbf{x}) \equiv \hat{z}_j(\mathbf{x}) \equiv \hat{y}_j(\mathbf{x}) \bmod \widehat{\mathfrak{m}}^\alpha.$$

We next put $z(\mathbf{x}) = (z_1(\mathbf{x}), \ldots, z_p(\mathbf{x}))$ and obtain

$$h_i(\mathbf{x}, z(\mathbf{x})) = Q_i(\mathbf{x}, y_{j\ell}(\mathbf{x}'))\Delta(\mathbf{x}, z(\mathbf{x})) \in \Delta(\mathbf{x}, z(\mathbf{x}))\mathbb{K}\{\mathbf{x}\},$$

which concludes the proof of C).

We can now complete the proof of the proposition. Consider new indeterminates $\mathbf{Y} = (Y_1, \ldots, Y_p)$ and the system

$$F_i(\mathbf{x}, \mathbf{Y}) = h_i(\mathbf{x}, \mathbf{Y} + z(\mathbf{x})) = 0, \ 1 \leq i \leq s. \qquad (3)$$

It holds

$$F_i(\mathbf{x}, 0) = h_i(\mathbf{x}, z(\mathbf{x})) \in x_n^\alpha \delta(\mathbf{x}, z(\mathbf{x}))^2 \mathbb{K}\{\mathbf{x}\}$$

and

$$\frac{D(F_1, \ldots, F_s)}{D(Y_1, \ldots, Y_s)}(\mathbf{x}, 0) = \delta(\mathbf{x}, z(\mathbf{x})),$$

so that by Tougeron's Implicit Functions Theorem (Proposition 1.3) we obtain a solution

$$Y_1(\mathbf{x}), \ldots, Y_p(\mathbf{x}) \in \mathbf{x}_n^\alpha \mathbb{K}\{\mathbf{x}\}$$

of the system (3) above. Then

$$y_i(\mathbf{x}) = Y_i(\mathbf{x}) + z(\mathbf{x}) = z(\mathbf{x}) \bmod \widehat{\mathfrak{m}}^\alpha,$$

and we set $y(\mathbf{x}) = (y_1(\mathbf{x}), \ldots, y_s(\mathbf{x}))$. Clearly, $y(\mathbf{x}) \in \mathbb{K}\{\mathbf{x}\}^p$ is the solution we were looking for, and the proof of the proposition finishes here. □

4 Formal Completion of Analytic Rings

In this section we obtain the basic comparison theorems for analytic rings and their formal completions. Proofs will consist of the systematic application of M. Artin's Approximation Theorem. In order to do so, we first state a slightly weaker version of that theorem.

Let $A = \mathbb{K}\{\mathbf{x}\}/I$ be an analytic ring, with $\mathbf{x} = (\mathbf{x}_1, \ldots, \mathbf{x}_n)$. Its *formal completion* is the formal ring $\widehat{A} = \mathbb{K}[[\mathbf{x}]]/I\mathbb{K}[[\mathbf{x}]]$. The extension of an ideal $\mathfrak{a} \subset A$ to the formal completion will be denoted by $\widehat{\mathfrak{a}} = \mathfrak{a}\widehat{A}$. Note that the extension of the maximal ideal $\mathfrak{m} = \mathfrak{m}_A$ is the maximal ideal $\widehat{\mathfrak{m}} = \mathfrak{m}_{\widehat{A}}$.

Proposition 4.1 *Let* $y = (y_1, \ldots, y_p)$ *be new indeterminates and* $P_1, \ldots, P_s \in A[y]$. *Let* $\hat{a}_1, \ldots, \hat{a}_p \in \widehat{A}$ *be a solution of the system*

$$P_1(y) = 0, \ldots, P_s(y) = 0.$$

Then, for every integer $\alpha \geq 1$ *there is a solution* $a_1, \ldots, a_p \in A$ *of the system such that*

$$a_1 = \hat{a}_1, \ldots, a_p = \hat{a}_p \bmod \widehat{\mathfrak{m}}^\alpha.$$

Proof. Let $f_1, \ldots, f_r \in \mathbb{K}\{\mathbf{x}\}$ be generators of I, $H_1, \ldots, H_s \in \mathbb{K}\{\mathbf{x}\}[y]$ polynomials whose classes mod I are $P_1, \ldots, P_s$, and $\hat{g}_1, \ldots, \hat{g}_p \in \mathbb{K}[[\mathbf{x}]]$ formal power series whose classes mod $I\mathbb{K}[[\mathbf{x}]]$ are $\hat{a}_1, \ldots, \hat{a}_p$. Then,

$$H_i(\mathbf{x}, \hat{g}_1(\mathbf{x}), \ldots, \hat{g}_p(\mathbf{x})) = \hat{\lambda}_{i1} f_1 + \cdots + \hat{\lambda}_{ir} f_r, \ 1 \leq i \leq s,$$

for some $\hat{\lambda}_{ij} \in \mathbb{K}[[\mathbf{x}]]$. Applying M. Artin's Approximation Theorem (Proposition 2.1) to the system

$$H_i(\mathbf{x}, y_1, \ldots, y_p) = z_{i1} f_1 + \cdots + z_{ir} f_r, \ 1 \leq i \leq s,$$

where the z_{ij}'s are new indeterminates, we get convergent power series $g_\ell, \lambda_{ij} \in \mathbb{K}\{\mathbf{x}\}$ such that

$$g_1 - \hat{g}_1, \ldots, g_1 - \hat{g}_p \in (\mathbf{x}_1, \ldots, \mathbf{x}_n)^\alpha,$$

and
$$H_i(x, g_1(x), \ldots, g_p(x)) = \lambda_{i1} f_1 + \cdots + \lambda_{ir} f_r, \quad 1 \leq i \leq s.$$
Consequently, the classes $a_\ell = g_\ell \bmod I$, $1 \leq \ell \leq s$ are the solution we sought. □

We can now deduce Nagata's comparison results. First of all:

Proposition 4.2 *(Flatness) We have:*

a) $\hat{\mathfrak{a}} \cap A = \mathfrak{a}$ *for every ideal* $\mathfrak{a}$ *of* A. *In particular, the canonical homomorphism* $A \to \hat{A}$ *is injective, and we write* $A \subset \hat{A}$.

b) $(\mathfrak{a} \cap \mathfrak{b})\widehat{\ } = \hat{\mathfrak{a}} \cap \hat{\mathfrak{b}}$ *for every two ideals* $\mathfrak{a}, \mathfrak{b}$ *of* A.

c) An element $\delta \in A$ *is a zero divisor in* $\hat{A}$ *if and only if it is a zero divisor in* A.

d) Let K *denote the total ring of fractions of* A. *Then* $\hat{A} \cap K = A$.

Proof. a) Let $a_1, \ldots, a_p$ be generators of $\mathfrak{a}$. If $c \in \hat{\mathfrak{a}} \cap A$, then there are $\hat{\lambda}_1, \ldots, \hat{\lambda}_p \in \hat{A}$ such that
$$c = \hat{\lambda}_1 a_1 + \cdots + \hat{\lambda}_p a_p.$$
Given each $\alpha \geq 1$ we can choose elements $\lambda_1, \ldots, \lambda_p \in A$ such that
$$\lambda_1 = \hat{\lambda}_1, \ldots, \lambda_p = \hat{\lambda}_p \bmod \hat{\mathfrak{m}}^\alpha.$$
Then,
$$c_\alpha = \lambda_1 a_1 + \cdots + \lambda_p a_p \in \mathfrak{a}, \quad c - c_\alpha \in \mathfrak{m}^\alpha,$$
and so,
$$c \in \bigcap_\alpha (\mathfrak{a} + \mathfrak{m}^\alpha) = \mathfrak{a}.$$

b) Let $a_1, \ldots, a_p$ be generators of $\mathfrak{a}$ and $b_1, \ldots, b_q$ generators of $\mathfrak{b}$. If $\hat{c} \in \hat{\mathfrak{a}} \cap \hat{\mathfrak{b}}$, we have
$$\hat{c} = \hat{\lambda}_1 a_1 + \cdots + \hat{\lambda}_p a_p = \hat{\mu}_1 b_1 + \cdots + \hat{\mu}_p b_p,$$
for some $\hat{\lambda}_i, \hat{\mu}_j \in \hat{A}$. We then consider the polynomial equation
$$y_1 a_1 + \cdots + y_p a_p = z_1 b_1 + \cdots + z_p b_p.$$
By Proposition 4.1, for each $\alpha \geq 1$ we get $\lambda_i, \mu_j \in A$ such that
$$\lambda_i = \hat{\lambda}_i, \ \mu_j = \hat{\mu}_j \bmod \hat{\mathfrak{m}}^\alpha,$$
and an element

$$c_\alpha = \lambda_1 a_1 + \cdots + \lambda_p a_p = \mu_1 b_1 + \cdots + \mu_p b_p.$$

Clearly

$$c_\alpha \in \mathfrak{a} \cap \mathfrak{b}, \quad \hat{c} = c_\alpha \bmod \widehat{\mathfrak{m}}^\alpha.$$

Consequently

$$\hat{c} \in \bigcap_\alpha \left(\widehat{(\mathfrak{a} \cap \mathfrak{b})} + \widehat{\mathfrak{m}}^\alpha \right) = \widehat{(\mathfrak{a} \cap \mathfrak{b})}.$$

This shows one inclusion, and the other is immediate.

c) Suppose there is a non-zero element $\hat{a} \in \hat{A}$ such that $\delta \hat{a} = 0$. Then, by Proposition 4.1, for every $\alpha \geq 1$ there is an element $a_\alpha \in A$ such that

$$\delta a_\alpha = 0, \quad a = a_\alpha \bmod \widehat{\mathfrak{m}}^\alpha.$$

If α is large enough, then $a_\alpha \neq 0$, and we see that δ is a zero divisor in A. The opposite implication is trivial.

d) Let $a, \delta \in A$, where δ is not a zero divisor in A, be such that $a/\delta \in \hat{A}$. This means that the equation

$$a = \delta \mathbf{y}$$

has some solution in $\hat{A}$, and by Proposition 4.1 it has one in A. We are done. □

The proofs required to get the preceding proposition only involve linear equations, which are the true concern of flatness, and there are much more direct proofs of the same facts without using the approximation technique. Our approach here is chosen so to avoid diversions and to be consistent with the other arguments of the section.

Proposition 4.3 *(Primary decompositions and dimension)* We have:

a) $\sqrt{\hat{\mathfrak{a}}} = \widehat{\sqrt{\mathfrak{a}}}$ for every ideal $\mathfrak{a} \subset A$.

b) The extension $\hat{\mathfrak{q}}$ of a primary ideal $\mathfrak{q} \subset A$ is a primary ideal.

c) The extension $\hat{\mathfrak{p}}$ of a prime ideal $\mathfrak{p} \subset A$ is a prime ideal, and $\mathrm{ht}(\mathfrak{p}) = \mathrm{ht}(\hat{\mathfrak{p}})$.

d) Let $\mathfrak{a}$ be an ideal of A, $\mathfrak{a} = \mathfrak{q}_1 \cap \cdots \cap \mathfrak{q}_r$ its primary decomposition, and $\mathfrak{p}_1 = \sqrt{\mathfrak{q}_1}, \ldots, \mathfrak{p}_r = \sqrt{\mathfrak{q}_r}$ its associated primes. Then, the primary decomposition of $\hat{\mathfrak{a}}$ is $\hat{\mathfrak{a}} = \hat{\mathfrak{q}}_1 \cap \cdots \cap \hat{\mathfrak{q}}_r$, and its associated primes are $\hat{\mathfrak{p}}_1, \ldots, \hat{\mathfrak{p}}_r$.

Proof. a) Let us prove the non-trivial inclusion. Consider any generators $a_1, \ldots, a_r$ of $\mathfrak{a}$. If $\hat{c} \in \sqrt{\hat{\mathfrak{a}}}$, then there are an integer $p \geq 1$ and elements $\hat{\lambda}_1, \ldots, \hat{\lambda}_r \in \hat{A}$ such that

$$\hat{c}^p = \hat{\lambda}_1 a_1 + \cdots + \hat{\lambda}_r f_r.$$

Once again by Proposition 4.1, for every $\alpha \geq 1$ we get $c_\alpha, \lambda_1, \ldots, \lambda_r \in A$ such that

$$c_\alpha^p = \lambda_1 a_1 + \cdots + \lambda_r a_r$$

and $\hat{c} = c_\alpha \bmod \widehat{\mathfrak{m}}^\alpha$. It follows that $c_\alpha \in \sqrt{\mathfrak{a}}$ and consequently
$$c \in \bigcap_\alpha \left(\widehat{\sqrt{\mathfrak{a}}} + \widehat{\mathfrak{m}}^\alpha\right) = \widehat{\sqrt{\mathfrak{a}}}.$$

b) Let $a_1, \ldots, a_q$ be generators of $\mathfrak{q}$ and $\hat{b}, \hat{c} \in \widehat{A}$ such that $\hat{b}\hat{c} \in \widehat{\mathfrak{q}}$. Then,
$$\hat{b}\hat{c} = \hat{\lambda}_1 a_1 + \cdots + \hat{\lambda}_q a_q$$
for some $\hat{\lambda}_i \in \widehat{A}$. By Proposition 4.1, there are $b_\alpha, c_\alpha, \lambda_{i\alpha} \in A$ such that
$$\hat{b} = b_\alpha, \quad \hat{c} = c_\alpha \bmod \widehat{\mathfrak{m}}^\alpha, \quad b_\alpha c_\alpha = \lambda_{1\alpha} a_1 + \cdots + \lambda_{q\alpha} a_q \in \mathfrak{q}.$$

There are now two possibilities:

- $b_\alpha \in \mathfrak{q}$ for infinitely many α's. Then
$$\hat{b} \in \bigcap_\alpha (\widehat{\mathfrak{q}} + \widehat{\mathfrak{m}}^\alpha) = \widehat{\mathfrak{q}}.$$

- $b_\alpha \notin \mathfrak{q}$ for all α large. Then, as $\mathfrak{q}$ is primary, $c_\alpha \in \sqrt{\mathfrak{q}}$ for all α large, and so
$$\hat{c} \in \bigcap_\alpha \left(\widehat{\sqrt{\mathfrak{q}}} + \widehat{\mathfrak{m}}^\alpha\right) = \widehat{\sqrt{\mathfrak{q}}} \subset \sqrt{\widehat{\mathfrak{q}}}.$$

This shows that $\widehat{\mathfrak{q}}$ is primary.

c) That $\widehat{\mathfrak{p}}$ is prime is an immediate consequence of *a)* and *b)*, and then we get the inequality $\operatorname{ht}(\mathfrak{p}) \leq \operatorname{ht}(\widehat{\mathfrak{p}})$. Conversely, if $\operatorname{ht}(\mathfrak{p}) = r$, there are r elements $a_1, \ldots, a_r \in A$ such that $\sqrt{\{a_1, \ldots, a_r\} A_\mathfrak{p}} = \mathfrak{p} A_\mathfrak{p}$ ([A-McD 11.14]). It follows easily that $\sqrt{\{a_1, \ldots, a_r\} \widehat{A}_{\widehat{\mathfrak{p}}}} = \widehat{\mathfrak{p}} \widehat{A}_{\widehat{\mathfrak{p}}}$. Hence $\operatorname{ht}(\widehat{\mathfrak{p}}) \leq r$ ([A-McD 11.14] again).

d) is a consequence of the previous parts of the proposition. □

Corollary 4.4 *(Nagata) An analytic ring is a reduced ring (resp. an integral domain) if and only if so is its formal completion.*

Proposition 4.5 *(Nagata) Let $\mathfrak{p}$ be a prime ideal of our analytic ring $A = \mathbb{K}\{\mathbf{x}\}/I$ and $\mathfrak{q}$ a prime ideal of its completion $\widehat{A} = \mathbb{K}[[\mathbf{x}]]/I\mathbb{K}[[\mathbf{x}]]$ such that $\mathfrak{q} \cap A = \mathfrak{p}$. Then $A_\mathfrak{p}$ is regular if and only if $\widehat{A}_\mathfrak{q}$ is regular.*

Proof. Throughout the proof the same letter will denote an ideal of A (resp. $\widehat{A}$) and its inverse image in $\mathbb{K}\{\mathbf{x}\}$ (resp. $\mathbb{K}[[\mathbf{x}]]$).

Suppose first that $A_\mathfrak{p} = \mathbb{K}\{\mathbf{x}\}_\mathfrak{p}/I\mathbb{K}\{\mathbf{x}\}_\mathfrak{p}$ is regular. By the Regularity Jacobian Criterion (Proposition II.4.3), it holds
$$\mathfrak{p} \not\supset J_s(I), \quad \operatorname{ht}(I\mathbb{K}\{\mathbf{x}\}_\mathfrak{p}) \leq s. \tag{1}$$

It follows immediately from the definitions (II.4.1) that $J_s(I\mathbb{K}[[\mathbf{x}]]) = J_s(I)\mathbb{K}[[\mathbf{x}]]$, and, by flatness (Proposition 4.2 a)), we deduce that $\mathbb{K}\{\mathbf{x}\} \cap J_s(I\mathbb{K}[[\mathbf{x}]]) = J_s(I)$. This, together with the hypothesis $\mathfrak{q} \cap \mathbb{K}\{\mathbf{x}\} = \mathfrak{p}$ and the first condition in (1), imply

$$\mathfrak{q} \not\supset J_s(I\mathbb{K}[[\mathbf{x}]]). \tag{2}$$

Now, by Proposition 4.3 c) and d) we have $\mathrm{ht}(I\mathbb{K}\{\mathbf{x}\}_{\mathfrak{p}}) = \mathrm{ht}(I\mathbb{K}[[\mathbf{x}]]_{\widehat{\mathfrak{p}}})$, and, since $\mathfrak{q} \supset \widehat{\mathfrak{p}}$, $\mathrm{ht}(I\mathbb{K}[[\mathbf{x}]]_{\widehat{\mathfrak{p}}}) \geq \mathrm{ht}(I\mathbb{K}[[\mathbf{x}]]_{\mathfrak{q}})$. Consequently,

$$\mathrm{ht}(I\mathbb{K}[[\mathbf{x}]]_{\mathfrak{q}}) \leq s. \tag{3}$$

Thus, from (2) and (3) we conclude that $\widehat{A}_{\mathfrak{q}}$ is regular (again by the Regularity Jacobian Criterion).

Suppose conversely that $\widehat{A}_{\mathfrak{q}}$ is regular. First of all, we notice that $\mathfrak{q}$ contains some minimal prime $\mathfrak{q}'$ of $\widehat{\mathfrak{p}} = \mathfrak{p}\widehat{A}$. Since $\widehat{A}_{\mathfrak{q}}$ is regular and $\mathfrak{q}' \subset \mathfrak{q}$ also $\widehat{A}'_{\mathfrak{q}}$ is regular (this follows immediately from the Regularity Jacobian Criterion). In other words, we may assume that $\mathfrak{q}$ is a minimal prime of $\widehat{\mathfrak{p}}$. By Proposition 4.3 c) and d), we deduce that $\mathfrak{p}$ generates the maximal ideal $\mathfrak{q}\widehat{A}_{\mathfrak{q}}$ of $\widehat{A}_{\mathfrak{q}}$ and $d = \dim(A_{\mathfrak{p}}) = \dim(\widehat{A}_{\mathfrak{q}})$. Now, since the latter ring is regular by hypothesis, its maximal ideal can be generated by exactly d elements, and we claim that these elements can be chosen in $\mathfrak{p}$.

Indeed, set $\mathfrak{n} = \mathfrak{q}\widehat{A}_{\mathfrak{q}}$ and $\kappa = \widehat{A}_{\mathfrak{q}}/\mathfrak{q}\widehat{A}_{\mathfrak{q}}$. Then, any given elements generate $\mathfrak{n}$ if and only if their classes mod $\mathfrak{n}^2$ generate $\mathfrak{n}/\mathfrak{n}^2$ as κ-linear space (Nakayama's Lemma). Since $\mathfrak{n}$ can be generated by d elements, it follows that the dimension of that linear space is $\leq d$, and so we can select d generators from any given generators. In particular, from $\mathfrak{p}$.

Once our claim is proved, let $a_1, \ldots, a_d \in \mathfrak{p}$ generate of $\mathfrak{q}\widehat{A}_{\mathfrak{q}}$, and let us show that they also generate $\mathfrak{p}A_{\mathfrak{p}}$. Consider any $b \in \mathfrak{p}A_{\mathfrak{p}}$. There are then $d, c_1, \ldots, c_d \in \widehat{A}$, $d \notin \mathfrak{q}$ such that

$$db = c_1 a_1 + \cdots + c_d a_d.$$

Now, by Proposition 4.1, for every $\alpha \geq 1$ we find $d_\alpha, c_{1\alpha}, \ldots, c_{d\alpha}$ such that

$$d - d_\alpha \in \widehat{\mathfrak{m}}^\alpha, \quad d_\alpha b = c_{1\alpha} a_1 + \cdots + c_{d\alpha} a_d.$$

If α is large enough, then $d_\alpha \notin \mathfrak{p}$, for otherwise

$$d \in \bigcap_\alpha (\widehat{\mathfrak{p}} + \widehat{\mathfrak{m}}^\alpha) \subset \mathfrak{q}.$$

Hence, b belongs to the ideal generated in $A_{\mathfrak{p}}$ by $a_1, \ldots, a_d$, as wanted.

Whence, $\mathfrak{p}A_{\mathfrak{p}}$ can be generated by d elements, and the local ring $A_{\mathfrak{p}}$ is regular. $\square$

Proposition 4.6 *(Nagata) Let A and $\widehat{A}$ be integral domains and denote by K and $\widehat{K}$ their quotient fields. Then K is algebraically closed in $\widehat{K}$.*

Proof. Consider a polynomial
$$P(t) = c_0 t^p + c_1 t^{p-1} + \cdots + c_p \in A[t]$$
and $\hat{a}, \hat{b} \in \widehat{A}$, $\hat{b} \neq 0$ such that $\hat{a}/\hat{b}$ is a root of P. It follows
$$c_0 \hat{a}^p + c_1 \hat{a}^{p-1}\hat{b} + \cdots + c_p \hat{b}^p = 0.$$
Applying as usual Proposition 4.1, we find $a_\alpha, b_\alpha \in A$ such that
$$\hat{a} = a_\alpha, \hat{b} = b_\alpha \bmod \widehat{\mathfrak{m}}^\alpha, \ P(a_\alpha/b_\alpha) = 0$$
(note that since $\hat{b} \neq 0$, also $b_\alpha \neq 0$ for α large). On the other hand, K is a field of characteristic zero, and so P has at most p roots. We can thus assume $a_\alpha/b_\alpha = a_0/b_0$. But
$$\hat{a}b_0 - \hat{b}a_0 = a_\alpha b_0 - b_\alpha a_0 \bmod \widehat{\mathfrak{m}}^\alpha,$$
and so
$$\hat{a}b_0 - \hat{b}a_0 \in \widehat{\mathfrak{m}}^\alpha$$
for all α. Hence $\hat{a}b_0 - \hat{b}a_0 = 0$ and $\hat{a}/\hat{b} = a_0/b_0 \in K$. □

Proposition 4.7 *(Nagata) An analytic ring A is normal if and only if so is its formal completion $\widehat{A}$.*

Proof. Since a normal analytic (resp. formal) ring is always an integral domain (Proposition III.2.4), in view of Corollary 4.4 we can assume that A and $\widehat{A}$ are integral domains. We denote by K the quotient field of A and by $\widehat{K}$ that of $\widehat{A}$.

We suppose first that A is normal. Consider $\hat{a}, \hat{b} \in \widehat{A}$, $\hat{b} \neq 0$, $\hat{c}_1, \ldots, \hat{c}_p \in \widehat{A}$ such that
$$(\hat{a}/\hat{b})^p + \hat{c}_1(\hat{a}/\hat{b})^{p-1} + \cdots + \hat{c}_p = 0.$$
We deduce
$$\hat{a}^p + \hat{c}_1 \hat{a}^{p-1}\hat{b} + \cdots + \hat{c}_p \hat{b}^p = 0.$$
As usual, we obtain $a_\alpha, b_\alpha, c_{i\alpha} \in A$ with
$$a_\alpha^p + c_{1\alpha} a_\alpha^{p-1} b_\alpha + \cdots + c_{p\alpha} b_\alpha^p = 0, \quad \hat{a} = a_\alpha, \hat{b} = b_\alpha \bmod \widehat{\mathfrak{m}}^\alpha$$
and since A is integrally closed in K, we deduce that $a_\alpha/b_\alpha \in A$, that is,
$$a_\alpha \in b_\alpha A.$$
It follows that

4 Formal Completion of Analytic Rings

$$a \in \bigcap_\alpha \left(a_\alpha \widehat{A} + \widehat{\mathfrak{m}}^\alpha\right) \subset \bigcap_\alpha \left(b_\alpha \widehat{A} + \widehat{\mathfrak{m}}^\alpha\right) \subset \bigcap_\alpha \left(\hat{b}\widehat{A} + \widehat{\mathfrak{m}}^\alpha\right) = \hat{b}\widehat{A},$$

and $\hat{a}/\hat{b} \in \widehat{A}$. Whence $\widehat{A}$ is integrally closed in $\widehat{K}$.

Conversely, if $\widehat{A}$ is normal, A is normal, too by Proposition 4.2 d). □

From the preceding proposition, the good behaviour of associated primes under completion (Proposition 4.3), and the description of normalizations given in Proposition III.2.3, we get:

Proposition 4.8 *Let A be a reduced analytic ring, $\widehat{A}$ its formal completion and $B = B_1 \times \cdots \times B_s$ its normalization. Then, the normalization of $\widehat{A}$ is the completion of B; that is,*

$$(\widehat{A})^\nu \simeq \widehat{B_1} \times \cdots \times \widehat{B_s}.$$

We conclude this section with some specific properties of the real case:

Proposition 4.9 *(Risler) Let $\mathbb{K} = \mathbb{R}$, and suppose that A and $\widehat{A}$ are integral domains. Then,*

a) A is 2-real if and only if so is $\widehat{A}$.

b) A is formally real if and only if so is $\widehat{A}$.

c) Every ordering of A extends to an ordering of $\widehat{A}$.

Proof. a) If $\widehat{A}$ is not 2-real, there are $\hat{a}, \hat{b} \in \widehat{A}$, neither of them zero, such that $\hat{a}^2 + \hat{b}^2 = 0$. Then for every $\alpha \geq 1$ there are $a_\alpha, b_\alpha \in A$ such that

$$a_\alpha^2 + b_\alpha^2 = 0; \quad \hat{a} = a_\alpha, \hat{b} = b_\alpha \bmod \widehat{\mathfrak{m}}^\alpha.$$

Thus, if α is large, neither a_α nor b_α are zero, and A is not 2-real.

b) can be proved as a), using Artin-Schreier's Criterion (proof of Proposition IV.4.1), but it also follows from c), proved below.

c) Let $M = \{a_i \mid i \in I\}$ be the collection of all elements of A wich are positive in a fixed ordering of A. We must show that there is an ordering of $\widehat{A}$ in which all the a_i's are positive. According to *Serre's Criterion* ([L XI.2 Cor to Th.4]) this follows if we show that no equation

$$a_{i_1}\mathbf{y}_1^2 + \cdots + a_{i_r}\mathbf{y}_r^2 = 0$$

has non-trivial solutions in $\widehat{A}$. But, were there a non-trivial solution in $\widehat{A}$, by the usual approximation trick there would be one in A, which is impossible, because all the a_i's are positive in the fixed ordering. □

5 Nash Rings

Let $\mathbb{K} = \mathbb{R}$ or $\mathbb{C}$.

(5.1) Algebraic power series. A power series $f \in \mathbb{K}[[\mathbf{x}]]$, where $\mathbf{x} = (\mathbf{x}_1, \ldots, \mathbf{x}_n)$, is called *algebraic* if there is a non-zero polynomial

$$P(\mathbf{x}, \mathbf{t}) = a_0(\mathbf{x})\mathbf{t}^p + a_1(\mathbf{x})\mathbf{t}^{p-1} + \cdots + a_p(\mathbf{x}) \in \mathbb{K}[\mathbf{x}, \mathbf{t}]$$

such that $P(\mathbf{x}, f(\mathbf{x})) = 0$.

We notice that an algebraic power series is convergent. Indeed, by M. Artin's Approximation Theorem, there are convergent power series $f_\alpha(\mathbf{x})$ whose Taylor expansions coincide with that of $f(\mathbf{x})$ up to an arbitrarily high order α and such that $P(\mathbf{x}, f_\alpha(\mathbf{x})) = 0$. If $f_\alpha \neq f$ for infinitely many α's, then the polynomial $P(\mathbf{x}, \mathbf{t})$ would have infinitely many roots, which is impossible. □

In other words, algebraic power series are the elements of $\mathbb{K}\{\mathbf{x}\}$ which are algebraic over $\mathbb{K}[\mathbf{x}]$. By the general properties of algebraic dependence ([L VII.1]), they form a ring which will be denoted by either $\mathcal{N}_n$, $\mathbb{K}\langle\mathbf{x}\rangle$, or $\mathbb{K}\langle\mathbf{x}_1, \ldots, \mathbf{x}_n\rangle$. If $f \in \mathbb{K}\langle\mathbf{x}\rangle$ and $f(0) \neq 0$, then $1/f \in \mathbb{K}[[\mathbf{x}]]$ is algebraic, and consequently f is a unit in $\mathbb{K}\langle\mathbf{x}\rangle$. Thus, the ring of algebraic power series is a local ring with maximal ideal

$$\{f \in \mathbb{K}\langle\mathbf{x}\rangle \mid f(0) = 0\},$$

and we have the local inclusions $\mathcal{N}_n \subset \mathcal{O}_n \subset \mathcal{F}_n$.

(5.2) Substitution. Let $\mathbf{y} = (\mathbf{y}_1, \ldots, \mathbf{y}_p)$ and $\mathbf{t}$ be new indeterminates, and $P \in \mathbb{K}[[\mathbf{y}]][\mathbf{t}]$ a distinguished polynomial of degree p. We denote by F the algebraic closure of the quotient field of $\mathbb{K}[[\mathbf{y}]]$ and consider a root $\xi \in F$ of P.

For $f(\mathbf{y}, \mathbf{t}) \in \mathbb{K}[[\mathbf{y}, \mathbf{t}]]$, let $R(\mathbf{y}, \mathbf{t}) \in \mathbb{K}[[\mathbf{y}]][\mathbf{t}]$ be the remainder of the formal division of f by P (Proposition I.3.2), and set

$$f(\mathbf{y}, \xi) = R(\mathbf{y}, \xi) \in F.$$

Then:

a) The map $f(\mathbf{y}, \mathbf{t}) \mapsto f(\mathbf{y}, \xi)$ is a homomorphism of $\mathbb{K}$-algebras.

Only the equality $(f_1 f_2)(\mathbf{y}, \xi) = f_1(\mathbf{y}, \xi) f_2(\mathbf{y}, \xi)$ needs some explanation. This equality is clear in case $f_1, f_2 \in \mathbb{K}[[\mathbf{y}]][\mathbf{t}]$, and the general case reduces to this replacing f_1, f_2 by their remainders after division by P. (Compare with the more general construction in the second half of the proof of Lemma III.1.1.) □

b) If $f \in \mathbb{K}\langle\mathbf{y}, \mathbf{t}\rangle$ and $f(\mathbf{y}, \xi) = 0$, then ξ is algebraic over $\mathbb{K}[\mathbf{y}]$.

Consider any equation

$$a_0(\mathbf{y}, \mathbf{t}) f^q(\mathbf{y}, \mathbf{t}) + \cdots + a_q(\mathbf{y}, \mathbf{t}) = 0, \quad a_0, \ldots, a_q \in \mathbb{K}[\mathbf{y}, \mathbf{t}], a_q \neq 0.$$

By *a)* we can substitute $\mathbf{t} = \xi$ to get $a_q(\mathbf{y}, \xi) = 0$, since $a_q(\mathbf{y}, \mathbf{t}) \neq 0$, we are done. □

5 Nash Rings

c) *If $f \in \mathbb{K}\langle \mathbf{y}, \mathbf{t} \rangle$ and ξ is algebraic over the polynomials, then $f(\mathbf{y}, \xi)$ is also algebraic over the polynomials.*

Consider again an equation
$$a_0(\mathbf{y}, \mathbf{t})f(\mathbf{y}, \mathbf{t})^q + \cdots + a_q(\mathbf{y}, \mathbf{t}) = 0, \ a_0, \ldots, a_q \in \mathbb{K}[\mathbf{y}, \mathbf{t}], a_q \neq 0,$$
with the additional condition that the coefficients $a_i(\mathbf{y}, \mathbf{t}) \in \mathbb{K}(\mathbf{y})[\mathbf{t}]$ have no common irreducible factor (which is possible because the ring $\mathbb{K}(\mathbf{y})[\mathbf{t}]$ is factorial, [L V.6 Th.10]). Making the substitution $\mathbf{t} = \xi$ we obtain
$$a_0(\mathbf{y}, \xi)f(\mathbf{y}, \xi)^q + \cdots + a_q(\mathbf{y}, \xi) = 0,$$
and we claim that some coefficient $a_i(\mathbf{y}, \xi) \neq 0$. Otherwise, all the $a_i(\mathbf{y}, \mathbf{t})$'s would be multiples of the irreducible polynomial of ξ over $\mathbb{K}(\mathbf{y})$, and they were chosen without common factors. □

Proposition 5.3 *Let $f, g_1, \ldots, g_n \in \mathcal{N}_n$, with $g_1(0) = \cdots = g_n(0) = 0$. Then, substitution gives an algebraic power series $f(g_1, \ldots, g_n) \in \mathcal{N}_n$.*

Proof. We first rename some variables, writing $g_1(\mathbf{z}), \ldots, g_n(\mathbf{z})$. We then consider the division
$$f(\mathbf{x}) = Q(\mathbf{z}, \mathbf{x})(\mathbf{x}_n - g_n(\mathbf{z})) + R(\mathbf{z}, \mathbf{x}_1, \ldots, \mathbf{x}_{n-1}),$$
where $R(\mathbf{z}, \mathbf{x}_1, \ldots, \mathbf{x}_{n-1}) \in \mathbb{K}[[\mathbf{x}_1, \ldots, \mathbf{x}_{n-1}]]$. Thus
$$f(\mathbf{x}_1, \ldots, \mathbf{x}_{n-1}, g_n(\mathbf{z})) = R(\mathbf{z}, \mathbf{x}_1, \ldots, \mathbf{x}_{n-1}),$$
and we get the same element that in 5.2 with
$$\mathbf{y} = (\mathbf{z}, \mathbf{x}_1, \ldots, \mathbf{x}_{n-1}), \ \mathbf{t} = \mathbf{x}_n, \ P = \mathbf{x}_n - g_n(\mathbf{z}), \ \xi = g_n(\mathbf{z}).$$
Hence, by 5.2 c),
$$f(\mathbf{x}_1, \ldots, \mathbf{x}_{n-1}, g_n(\mathbf{z})) \in \mathbb{K}\langle \mathbf{z}, \mathbf{x}_1, \ldots, \mathbf{x}_{n-1} \rangle.$$
The proof ends by repeating this argument $n - 1$ more times. □

Proposition 5.4 *The derivatives of an algebraic power series are algebraic power series, too.*

Proof. Let $f \in \mathcal{N}_n$, and
$$P(\mathbf{x}, f(\mathbf{x})) = a_0(\mathbf{x})f(\mathbf{x})^p + \cdots + a_p(\mathbf{x}) = 0$$
be an equation of algebraic dependence of f. We then have

$$0 = \frac{\partial}{\partial x_i}(P(x, f(x))) = \sum_{j=0}^{p} \frac{\partial a_j}{\partial x_i} f^{p-j} + \frac{\partial f}{\partial x_i} \sum_{j=0}^{p-1} a_j f^{p-j-1}.$$

If we choose P of minimal degree, the second sum in the right hand side of this equality is not zero, and since the derivatives of the a_j's are polynomials, the power series $\partial f/\partial x_i$ is algebraic over $\mathbb{K}(x)[f]$, and consequently over $\mathbb{K}(x)$. □

Remarks 5.5 *a)* Every algebraic power series $f \in \mathbb{K}\langle x \rangle$ can be written in the form

$$f = f(0) + x_1 f_1 + \cdots + x_n f_n,$$

where $f_i \in \mathbb{K}\langle x_1, \ldots, x_i \rangle$ for $1 \leq i \leq n$.

We obtain

$x_1 f_1$ by the substitution $x_2 = \cdots = x_n = 0$ in $f - f(0)$,

$x_2 f_2$ by the substitution $x_3 = \cdots = x_n = 0$ in $f - f(0) - x_1 f_1$,

and so forth. □

b) The indeterminates $x_1, \ldots, x_n$ generate the maximal ideal of $\mathcal{N}_n$.

In fact, the maximal ideal consists of the f's such that $f(0) = 0$. □

c) Let $f = \sum_{p \geq 0} a_p(x') x_n \in \mathbb{K}[[x'x_n]]$, where $x' = (x_1, \ldots, x_{n-1})$. If f is algebraic, then all the coefficients a_p are algebraic, since

$$a_p(x') = \frac{1}{p!} \frac{\partial f}{\partial x_n}(x', 0).$$

□

The essential fact is the following:

Proposition 5.6 *(Rückert's Division Theorem for algebraic power series) Proposition I.3.2 is also valid when substituting $\mathcal{O}_n$ by $\mathcal{N}_n$ and $\mathcal{O}_{n-1}$ by $\mathcal{N}_{n-1}$.*

Proof. Let $\Phi \in \mathcal{N}_n$ be regular of order p with respect to x_n and $f \in \mathcal{N}_n$. By the Preparation Theorem (Proposition I.3.3) and Remark II.2.2, we have

$$\Phi = U P_1 \cdots P_r, \quad U \in \mathcal{F}_n, \quad U(0) \neq 0,$$

where each P_i is an irreducible distinguished polynomial of $\mathcal{F}_{n-1}[x_n]$ of degree p_i. The P_i's are not necessarily distinct, and $p = p_1 + \cdots + p_r$. Let $x' = (x_1, \ldots, x_{n-1})$.

We can apply the substitution procedure described in 5.2 to every root ξ of every P_i, so that $\Phi(x', \xi) = 0$, and by 5.2 *b)*, ξ is algebraic over $\mathbb{K}[x']$. But the coefficients of the polynomial P_i are the symmetric functions of its roots, and so those coefficients are algebraic over $\mathbb{K}[x']$. Consequently, $P_i \in \mathcal{N}_{n-1}[x_n]$ for all i and then $U \in \mathcal{N}_n$.

Now we notice that succesive divisions by $P_1, \ldots, P_r$ give the division by $U^{-1}\Phi$, hence by Φ. By all of this, we can suppose without loss of generality that Φ is irreducible.

Consider the formal division

$$f = Q\Phi + a_1(\mathbf{x}')x_n^{p-1} + \cdots + a_p(\mathbf{x}').$$

We must see that Q and the a_i's are algebraic, and for this it is enough that the a_i's are algebraic. To that end, we will use the p roots $\xi_1, \ldots, \xi_p$ of Φ in the algebraic closure F of the quotient field of $\mathcal{F}_{n-1}$. These roots are all different because Φ is an irreducible polynomial. Then, substituting as explained in 5.2, we get

$$f(\mathbf{x}', \xi_i) = a_1(\mathbf{x}')\xi_i^{p-1} + \cdots + a_p(\mathbf{x}'), \ 1 \le i \le p.$$

This linear system gives the a_ℓ's as rational functions of the ξ_i's and the $f(\mathbf{x}', \xi_j)$'s, since the determinant is

$$\prod_{i<j}(\xi_i - \xi_j) \ne 0.$$

Finally, by 5.2 c), the $f(\mathbf{x}', \xi_j)$'s are algebraic over $\mathbb{K}[\mathbf{x}']$ as the ξ_i's are, and we conclude that the coefficients a_ℓ's are algebraic as wanted. We are done. □

Once Rückert's Division Theorem is available, all the subsequent theory can be developed for algebraic power series as it was for convergent and formal power series. In doing this we introduce the following terminology:

Definition 5.7 *A* Nash ring *over* $\mathbb{K}$ *is a ring isomorphic to* $\mathbb{K}\langle\mathbf{x}\rangle/I$ *with* $\mathbf{x} = (\mathbf{x}_1, \ldots, \mathbf{x}_n)$; *usually we will not specify "over* $\mathbb{K}$". *If* A, B *are two Nash rings, a* Nash homomorphism $A \to B$ *is a homomorphism of* $\mathbb{K}$-*algebras. The field* $\mathbb{K}$ *is called the* coefficient field.

As said before, the full theory developed in the analytic category is valid in the *Nash category*: Weierstrass's Preparation Theorem, Hensel's Lemma, Mather's Finiteness Theorem, transversal changes of coordinates, dimension, the Local Parametrization Theorem, Nagata's Jacobian Criteria, complexification, Nullstellensätze, M. Artin's Approximation Theorem, completions. With respect to the latter, note that Nash rings have *formal completions*, as analytic rings have, but also *analytic completions*, defined in the obvious way. Of course the behaviour is always the same. We will use freely all these facts, quoting the corresponding result in the analytic category.

VI Local Algebraic Rings

Summary. We study in this chapter some local properties of real and complex algebraic varieties. This study consists of the comparison of the so-called local algebraic rings and their completions. In Section 1 we define the local algebraic rings and their Nash, analytic and formal completions, and check the typical flat behaviour. In Section 2 we prove Chevalley's Theorem concerning completions of local algebraic domains. Section 3 is devoted to Zariski's Main Theorem stating that the completion of a local algebraic normal domain is normal. In Section 4 we describe the completion of the normalization of a local algebraic domain. Finally, we obtain in Section 5 Efroymson's Theorem, which deals with the implications of adding reality assumptions to Chevalley's and Zariski's statements.

1 Local Algebraic Rings

We introduce in this section several local rings attached to a point of an algebraic subset of the affine space. These local rings are the suitable tools to study and compare algebraic and analytic properties at a given point. Clearly, we can assume without loss of generality that the point is the origin, and we will do so from now on.

Let $\mathbb{K} = \mathbb{R}$ or $\mathbb{C}$, and consider indeterminates $\mathbf{x} = (x_1, \ldots, x_n)$. We denote by $\mathcal{R}_n$, $\mathbb{K}_0[\mathbf{x}]$ or $\mathbb{K}_0[x_1, \ldots, x_n]$ the localization of the polynomial ring $\mathbb{K}[\mathbf{x}]$ at the maximal ideal generated by $x_1, \ldots, x_n$. Clearly

$$\mathcal{R}_n = \{f/g \mid f, g \in \mathbb{K}[\mathbf{x}],\ g(0) \neq 0\}.$$

This is a local regular ring of dimension n with maximal ideal

$$\{h \in \mathcal{R}_n \mid h(0) \neq 0\},$$

and the canonical inclusions

$$\mathcal{R}_n \subset \mathcal{N}_n \subset \mathcal{O}_n \subset \mathcal{F}_n$$

are local homomorphisms.

Definition 1.1 *A local algebraic ring over* $\mathbb{K}$ *is a ring isomorphic to* $\mathbb{K}_0[\mathbf{x}]/I$ *with* $\mathbf{x} = (x_1, \ldots, x_n)$; *usually we will not specify "over* $\mathbb{K}$*". If* A, B *are two local algebraic rings, a* local algebraic homomorphism $A \to B$ *is a homomorphism of* $\mathbb{K}$*-algebras. The field* $\mathbb{K}$ *is called the* coefficient *field.*

1 Local Algebraic Rings

Local algebraic rings are the local rings of the *algebraic category*. We have to consider the local rings corresponding to the other three categories. This is done as follows.

Let $A = \mathcal{R}_n/I$ be a local algebraic ring. The *Nash completion* of A is the Nash ring $\mathcal{N}_n/I\mathcal{N}_n$. Similarly, the *analytic completion* of A is $\mathcal{O}_n/I\mathcal{O}_n$, and the *formal completion* of A is $\mathcal{F}_n/I\mathcal{F}_n$.

Now let A^* be a completion of A (either Nash, analytic or formal). The extension to A^* of an ideal $\mathfrak{a}$ of A will be denoted by $\mathfrak{a}^*$. If $\mathfrak{m}$ is the maximal ideal of A, then $\mathfrak{m}^*$ is the maximal ideal of A^*.

After this preparation, comparing algebraic with Nash, analytic or formal properties is comparing A with its Nash, analytic or formal completion.

To study an extension $A \to A^*$ we do not dispose of M. Artin's Approximation Theorem, and so all comparison results of Section V.4 have to be revised. However we still have the following useful fact:

Lemma 1.2 *Let* $\mathbf{y} = (y_1, \ldots, y_p)$ *be new indeterminates and consider a linear polynomial*
$$P(\mathbf{y}) = c_0 + c_1 y_1 + \cdots + c_p y_p \in A[\mathbf{y}].$$
Let $a^* = (a_1^*, \ldots, a_p^*) \in A^{*p}$ *be a solution of the linear equation* $P(\mathbf{y}) = 0$. *Then for every integer* $\alpha \geq 1$ *there is a solution* $a = (a_1, \ldots, a_p) \in A^p$ *of the same equation such that*
$$a_1 = a_1^*, \ldots, a_1 = a_1^* \mod \mathfrak{m}^{*\alpha}.$$

Proof. Recall that $A = \mathcal{R}_n/I$, and pick generators $h_1, \ldots, h_q \in \mathcal{R}_n$ of the ideal I. Choose also elements $g_0, \ldots, g_p \in \mathcal{R}_n$ whose classes mod I are $c_0, \ldots, c_p$. Then, the linear equation given over A reduces to the following one over $\mathcal{R}_n$:
$$g_0 + g_1 y_1 + \cdots + g_p y_p + h_1 z_1 + \cdots + h_q z_q = 0,$$
where $z_1, \ldots, z_q$ are additional variables. Consequently, we can assume without loss of generality that $A = \mathcal{R}_n$.

Now, for every integer $\beta > \alpha$ set
$$a^* = a_\alpha + a' + a'',$$
where
$$a_{\alpha i} = \text{terms of degree } < \alpha \text{ of } a_i,$$
$$a_i' = \text{terms of degree } \geq \alpha \text{ and } < \beta \text{ of } a_i,$$
$$a_i'' = \text{terms of degree } \geq \beta \text{ of } a_i.$$

We have
$$a_{\alpha i}, a_i' \in \mathbb{K}[\mathbf{x}], \ a_i'' \in \mathcal{R}_n^*,$$
and
$$a_i' \in \mathfrak{m}^\alpha, \ a_i'' \in \mathfrak{m}^{*\beta}.$$

As $P(a^*) = 0$, we have

$$P(a_\alpha) + \sum_{i=1}^{p} c_i a_i' = -\sum_{i=1}^{p} c_i a_i'' \in \mathfrak{m}^{*\beta}.$$

Hence, the polynomial in the left hand side has no monomial of degree $< \beta$, that is, it belongs to $\mathfrak{m}^\beta$. This is valid for all $\beta > \alpha$ with α fixed, which by Krull's Theorem implies:

$$P(a_\alpha) \in \bigcap_\beta (\{c_1, \ldots, c_p\}\mathfrak{m}^\alpha + \mathfrak{m}^\beta) = \{c_1, \ldots, c_p\}\mathfrak{m}^\alpha.$$

Thus

$$P(a_\alpha) = \sum_{i=1}^{p} c_i b_i,$$

with $b_1, \ldots, b_p \in \mathfrak{m}^\alpha$, and finally, since $a_{\alpha i} - b_i \equiv a_{\alpha i} \equiv a_i^*$ mod $\mathfrak{m}^{*\alpha}$,

$$a = (a_{\alpha 1} - b_1, \ldots, a_{\alpha p} - b_p)$$

is the solution we were looking for. □

Once the preceding lemma is available, we can repeat the proof of Proposition V.4.2 to get:

Proposition 1.3 *(Flatness) We have:*

a) $\mathfrak{a}^* \cap A = \mathfrak{a}$ *for every ideal* $\mathfrak{a}$ *of* A*. In particular, the canonical homomorphism* $A \to A^*$ *is injective, and we will write* $A \subset A^*$*.*

b) $(\mathfrak{a} \cap \mathfrak{b})^* = \mathfrak{a}^* \cap \mathfrak{b}^*$ *for every two ideals* $\mathfrak{a}, \mathfrak{b}$ *of* A*.*

c) *An element* $\delta \in A$ *is a zero divisor in* A^* *if and only if it is a zero divisor in* A*.*

d) *Let* K *denote the total ring of fractions of* A*. Then* $A^* \cap K = A$*.*

2 Chevalley's Theorem

In this section we discuss primary decompositions, which is a more involved matter. We will show:

Proposition 2.1 *(Chevalley) Let* A *be a local algebraic ring and* A^* *a completion of* A*. Let* $\mathfrak{p} \subset A$ *be a prime ideal of height* r*. Then the extension* $\mathfrak{p}^* = \mathfrak{p}A^*$ *is a radical ideal whose associated prime ideals all have height* r*.*

In this statement A^* is either the Nash, the analytic or the formal completion. From it we get

Corollary 2.2 $\sqrt{\mathfrak{a}^*} = (\sqrt{\mathfrak{a}})^*$ *for every ideal* $\mathfrak{a} \subset A$. *In particular, A is reduced if and only if so is A^*.*

Proof. Clearly
$$\mathfrak{a}^* \subset (\sqrt{\mathfrak{a}})^* \subset \sqrt{\mathfrak{a}^*},$$
and we have to see that the ideal $(\sqrt{\mathfrak{a}})^*$ is radical. But $\sqrt{\mathfrak{a}}$ is radical, and so $\sqrt{\mathfrak{a}} = \mathfrak{p}_1 \cap \cdots \cap \mathfrak{p}_r$ for certain prime ideals $\mathfrak{p}_i$. By the proposition, the extensions $\mathfrak{p}_i^*$ are radical ideals, and by Proposition 1.3 b), we have
$$(\sqrt{\mathfrak{a}})^* = \mathfrak{p}_1^* \cap \cdots \cap \mathfrak{p}_r^*.$$
Hence $(\sqrt{\mathfrak{a}})^*$ is radical, and the proof is finished. □

Now, by Proposition V.4.3, it is enough to prove the proposition for the Nash completion. The advantage of Nash completions is the following:

Lemma 2.3 *Let A be a local algebraic ring and A^* its Nash completion.*

a) Let $\mathfrak{q} \subset A^*$ be a prime ideal, $\mathfrak{p} = \mathfrak{q} \cap A$ and K the quotient field of $A/\mathfrak{p}$. Then
$$\dim(A^*/\mathfrak{q}) = \mathrm{tr.deg.}(K : \mathbb{K}).$$

b) Let $\mathfrak{q} \subset \mathfrak{q}'$ be two prime ideals of A^*. If $\mathfrak{q} \cap A = \mathfrak{q}' \cap A$ then $\mathfrak{q} = \mathfrak{q}'$.

Proof. a) Let L be the quotient field of $A^*/\mathfrak{q}$. Then, by Noether's Projection Lemma (Proposition II.2.6), after a linear change of coordinates the canonical homomorphism $\mathbb{K}\langle x_1, \ldots, x_d \rangle$ is injective and finite, where $d = \dim(A^*/\mathfrak{q})$. Then, L is algebraic over $\mathcal{N}_d$, which in turn is algebraic over $\mathbb{K}[x_1, \ldots, x_d]$, and so
$$d = \mathrm{tr.deg.}(L : \mathbb{K}).$$
But $\mathbb{K}(x_1, \ldots, x_d) \subset K \subset L$, and consequently
$$d = \mathrm{tr.deg.}(K : \mathbb{K}).$$

b) By a), $\dim(A^*/\mathfrak{q}) = \dim(A^*/\mathfrak{q}')$, and since $\mathfrak{q} \subset \mathfrak{q}'$, it must be $\mathfrak{q} = \mathfrak{q}'$. □

We will now prove Proposition 2.1 for the Nash completion. In fact, we will prove the following quite stronger assertion:

Proposition 2.4 *Let A be a local algebraic ring and A^* its Nash completion. Let $\mathfrak{p} \subset A$ be a prime ideal of height r. Then, the extension $\mathfrak{p}^* = \mathfrak{p}A^*$ is a radical ideal whose associated prime ideals are exactly the prime ideals lying over $\mathfrak{p}$, and all of them have height r.*

Proof. Let $\mathfrak{q}$ be an associated prime of $\mathfrak{p}^*$. We will show that $\mathfrak{q} \cap A = \mathfrak{p}$.

By Proposition 1.3 *a)*, $\mathfrak{p} = \mathfrak{p}^* \cap A \subset \mathfrak{q} \cap A$. For the converse inclusion consider $\delta \in \mathfrak{q} \cap A$. Then δ is a zero divisor in the Nash ring $A^*/\mathfrak{p}^*$ ([A-McD 4.7]), which is the Nash completion of the local algebraic ring $A/\mathfrak{p}$. By Proposition 1.3 *c)*, δ is also a zero divisor in $A/\mathfrak{p}$ and so $\delta \in \mathfrak{p}$. This gives $\mathfrak{p} \supset \mathfrak{q} \cap A$.

Let now $\mathfrak{q}$ be a prime ideal lying over $\mathfrak{p}$: $\mathfrak{q} \cap A = \mathfrak{p}$. Then $\mathfrak{q} \supset \mathfrak{p}^*$, and consequently there is some associated prime $\mathfrak{q}'$ of $\mathfrak{p}^*$ contained in $\mathfrak{q}$. Thus, $\mathfrak{q}' \cap A = \mathfrak{p}$, and by Lemma 2.3 *b)* we get $\mathfrak{q}' = \mathfrak{q}$, and $\mathfrak{q}$ is an associated prime of $\mathfrak{p}^*$.

So far, we have shown that the associated primes of $\mathfrak{p}^*$ are the prime ideals $\mathfrak{q}$ of A^* lying over $\mathfrak{p}$. We will next see that all of them have height $r = \mathrm{ht}(\mathfrak{p})$.

First we notice that any chain of prime ideals of A^* contained in $\mathfrak{q}$ gives a chain of prime ideals of A contained in $\mathfrak{p}$, and by Lemma 2.3 the length of both chains is the same. Hence $r = \mathrm{ht}(\mathfrak{p}) \geq \mathrm{ht}(\mathfrak{q})$. Conversely, let $\mathfrak{p} = \mathfrak{p}_0 \supset \mathfrak{p}_1 \cdots \supset \mathfrak{p}_r$ be a chain of prime ideals of A. We then put $\mathfrak{q} = \mathfrak{q}_0$, and so $\mathfrak{q}_0 \supset \mathfrak{p}_1^*$. It follows that $\mathfrak{q}_0$ contains some associated prime $\mathfrak{q}_1$ of $\mathfrak{p}_1^*$, and we know that $\mathfrak{q}_1 \cap A = \mathfrak{p}_1$. Repeating this, we end up with a length r chain of prime ideals of A^* contained in $\mathfrak{q}$, and so $\mathrm{ht}(\mathfrak{q}) \geq r$.

It remains to see that $\mathfrak{p}^*$ is a radical ideal. To do so we can assume $A = \mathcal{R}_n$, $A^* = \mathcal{N}_n$ and use the Equidimensionality Jacobian Criterion (Proposition II.4.8): we will find an element in the regularity ideal $R_r(\mathfrak{p}^*)$ which is not a zero divisor in $A^*/\mathfrak{p}^*$, where $r = n - \dim(A^*/\mathfrak{p}^*) = \mathrm{ht}(\mathfrak{p}^*)$. We recall that those zero divisors are the elements of the associated primes $\mathfrak{q}$ of $\mathfrak{p}^*$, and it will be enough to see that no $\mathfrak{q}$ contains $R_r(\mathfrak{p}^*)$. This will finally follow from showing that $\mathfrak{p}^*$ generates the maximal ideal of the localization $A_\mathfrak{q}^*$, since in this case

$$A_\mathfrak{q}^*/\mathfrak{p}^* A_\mathfrak{q}^*$$

is a field, that is, a regular ring of dimension $0 = \mathrm{ht}(\mathfrak{q}) - r$, and by the Regularity Jacobian Criterion (Proposition II.4.3) $R_r(\mathfrak{p}^*) \not\subset \mathfrak{q}$. Let us prove, hence, that

$$\mathfrak{p}^* A_\mathfrak{q}^* = \mathfrak{q} A_\mathfrak{q}^*.$$

First of all, by Lemma 2.3 *a)*

$$\mathrm{tr.deg.}(K : \mathbb{K}) = \dim(A^*/\mathfrak{q}) = n - r,$$

where as usual K stands for the quotient field of $A/\mathfrak{p}$. We then apply Noether's Projection Lemma for polynomials ([A-McD §5 Ex.16]), and after a linear change of coordinates the classes $\mathbf{x}_i$ mod $\mathfrak{p}$, $1 \leq i \leq n - r$ are integral over the polynomial ring $\mathbb{K}[\mathbf{x}_1, \ldots, \mathbf{x}_{n-r}]$. Then, any equation of integral dependence of $\mathbf{x}_i$ mod $\mathfrak{p}$ gives a polynomial $P_i(\mathbf{x}', \mathbf{x}_i) \in \mathfrak{p} \cap \mathcal{R}_{n-r}[\mathbf{x}_i]$ and, maybe after substituting P_i by a derivative, we can suppose that $\partial P_i/\partial \mathbf{x}_i \notin \mathfrak{p}$ $(n - r < i \leq n)$. Clearly

$$\delta = \frac{D(P_{n-r+1}, \ldots, P_n)}{D(\mathbf{x}_{n-r+1}, \ldots, \mathbf{x}_n)} = \prod_{n-r < i \leq n} \frac{\partial P_i}{\partial \mathbf{x}_i} \notin \mathfrak{p} = \mathfrak{q} \cap \mathcal{R}_n.$$

Consider the ideal

$$I = \{P_{n-r+1}, \ldots, P_n\}A^* \subset \mathfrak{p}^*.$$

Since I is generated by r elements, $1 \in G_r(I')$ and the regularity ideal is $R_r(I) = \sqrt{J_r(I)}$. Hence, $\delta \in R_r(I)\setminus\mathfrak{q}$, and by the Regularity Jacobian Criterion (Proposition II.4.3) the local ring

$$A_\mathfrak{q}^*/IA_\mathfrak{q}^*$$

is regular of dimension $\text{ht}(\mathfrak{q}) - r = 0$. In other words,

$$\mathfrak{q}A_\mathfrak{q}^* = IA_\mathfrak{q}^* \subset \mathfrak{p}^*A_\mathfrak{q}^*.$$

□

We finally have:

Proposition 2.5 *Let A be a local algebraic ring and A^* a completion of A. Let $\mathfrak{p} \subset A$, $\mathfrak{q} \subset A^*$ be prime ideals such that $\mathfrak{q} \cap A = \mathfrak{p}$. Then $A_\mathfrak{p}$ is regular if and only if so is $A_\mathfrak{q}^*$.*

Proof. By Proposition V.4.5, we can assume without loss of generality that A^* is the Nash completion of A. Then $\mathfrak{q}$ is an associated prime of the extension $\mathfrak{p}^* = \mathfrak{q}A^*$, which is a radical ideal, and $d = \dim(A_\mathfrak{p}) = \dim(A_\mathfrak{q}^*)$. Hence, if $A_\mathfrak{p}$ is regular, its maximal ideal $\mathfrak{p}A_\mathfrak{p}$ is generated by d elements, which in turn generate $\mathfrak{p}A_\mathfrak{q}^* = \mathfrak{q}A_\mathfrak{q}^*$, and $A_\mathfrak{q}^*$ is regular. The converse implication follows by the same argument as in Proposition V.4.5, using Lemma 1.2 instead of Proposition V.4.1. □

3 Zariski's Main Theorem

In this section we prove the famous Zariski's Main Theorem. To start with, we need a lemma that mixes completions with the local parametrization techniques of Sections II.2, II.3 and II.4:

Lemma 3.1 *(Zariski's Condition D) Let A be a local algebraic domain, A^* a completion of A and B^* the normalization of A^*. There is then an element $\delta \in A$ such that $\delta B^* \subset A^*$.*

Proof. We will prove the result for the Nash completion, and the same argument would work for the other completions. First of all, let $A = \mathcal{R}_n/\mathfrak{p}$, where $\mathfrak{p}$ is a prime ideal of height, say, r of $\mathcal{R}_n$. Then $A^* = \mathcal{N}_n/\mathfrak{p}^*$ and, by Chevalley's Theorem (Proposition 2.4),

$$\mathfrak{p}^* = \mathfrak{q}_1 \cap \cdots \cap \mathfrak{q}_s,$$

where the $\mathfrak{q}_i$ have height r and lie over $\mathfrak{p}$. By notational convenience we denote also by $\mathfrak{q}_i$ the associated prime $\mathfrak{q}_i \mod \mathfrak{p}^*$ of $(0) \subset A^*$. We recall that

$$d = n - r = \text{tr.deg.}(K : \mathbb{K}),$$

where K stands for the quotient field of A (Lemma 2.3 a)).

Next, we recall the description of the normalization B^* of A^* given in Proposition III.2.2. Let K^* denote the total ring of fractions of A^* and K_i^* that of $A_i^* = \mathcal{N}_n/\mathfrak{q}_i$ for $1 \leq i \leq s$. Then K^* is canonically isomorphic to $K_1^* \times \cdots \times K_s^*$, and this isomorphism maps B^* onto the product $B_1^* \times \cdots \times B_s^*$, where the B_i^*'s are the normalizations of the A_i^*'s.

We now apply simoultaneously Noether's Projection Lemma to $\mathfrak{p}$ and the $\mathfrak{q}_i$'s. This is possible because the linear changes used for polynomials, and the ones used for power series are of the same type (compare [A-McD 5 Ex.16] with II.2.3 and Proposition II.2.6). In the end we have the following situation:

- Every canonical homomorphism $\mathcal{N}_d \to A_i^*$ is finite and injective, and the element $\theta_i = \mathbf{x}_{d+1} \bmod \mathfrak{q}_i$ is a primitive element of K_i^* over the quotient field L^* of $\mathcal{N}_d$.

- The canonical homomorphism $\mathcal{R}_d \to A$ is injective, and the elements
$$\theta = \mathbf{x}_{d+1} \bmod \mathfrak{p}, \mathbf{x}_{d+2} \bmod \mathfrak{p}, \ldots, \mathbf{x}_n \bmod \mathfrak{p}$$
are integral over $\mathcal{R}_d$.

We are thus in the conditions of Propositions II.3.2 and II.4.4, from which we stress:

- The irreducible polynomial of θ_i over L^* is a distinguished polynomial $P_i \in \mathcal{N}_d[\mathbf{t}]$ whose discriminant $\delta_i \in \mathcal{N}_d$ has the property that $\delta_i B_i^* \subset A_i^*$.

On the other hand, θ is integral over $\mathcal{R}_d$ and so there is a monic polynomial $P \in \mathcal{R}_n[\mathbf{t}]$ such that $P(\theta) = 0$. This implies $P(\theta_i) = 0$ for each i, and since P_i is the irreducible polynomial of θ_i and it is distinguished, we have $P = UP_i$ for some $U \in \mathcal{N}_d[\mathbf{t}]$. Consequently, $\delta = \eta \delta_i$, where $\delta \in \mathcal{R}_n$ is the discriminant of P and $\eta \in \mathcal{N}_d$ (by [L V.10] this η is exactly the product of the discriminant of U and the resultant of U, P_i, but we do not need such a detail here). Hence, from the last remark we deduce:

a) There is an element $\delta \in \mathcal{R}_d \subset A \subset A^*$ such that $\delta B_i^* \subset A_i^*$ for all i.

We next use that every $\mathbf{x}_i \bmod \mathfrak{p}$ ($d+1 < i < n$) is integral over $\mathcal{R}_d$: we pick equations of integral dependence

$$Q_i(\mathbf{x}_i \bmod \mathfrak{p}) = 0$$

with $Q_i \in \mathcal{R}_d[\mathbf{t}]$ of minimal degree. Furthermore, we put $Q_{d+1} = P$, and consider the Jacobian

$$\delta' = \frac{D(Q_{d+1}(\mathbf{x}_{d+1}), \ldots, Q_n(\mathbf{x}_n))}{D(\mathbf{x}_{d+1}, \ldots, \mathbf{x}_n)} = \frac{\partial Q_{d+1}(\mathbf{x}_{d+1})}{\partial \mathbf{x}_{d+1}} \cdots \frac{\partial Q_n(\mathbf{x}_n)}{\partial \mathbf{x}_n} \in \mathcal{R}_n.$$

By the minimality of the Q_i's, this Jacobian does not belong to $\mathfrak{p}$, and consequently $\delta' \notin \mathfrak{p}^*$. Since $\delta' \in J_r(\mathfrak{p}^*)$ and $\mathfrak{p}^*$ is a reduced ideal whose associated primes have all height r, we deduce from the Regularity Jacobian Criterion (Proposition II.4.3):

3 Zariski's Main Theorem

b) *The localization $A_{\mathfrak{q}}^*$ is a local regular ring for every prime ideal $\mathfrak{q}$ such that $\delta' \notin \mathfrak{q}$.*

After this preparation, we will prove the lemma by showing that

$$(\delta\delta')^p B^* \subset A^*$$

for some integer $p \geq 0$.

To see this, we put $\rho = \delta\delta'$ and remark that from *a)* it follows

$$\rho B^* \subset (\rho B_1^*) \times \cdots \times (\rho B_s^*) \subset A_1^* \times \cdots \times A_s^*.$$

Thus, it is enough to see that

$$\rho^q (A_1^* \times \cdots \times A_s^*) \subset A^*$$

for some $q \geq 0$. Furthermore, $A_1^* \times \cdots \times A_s^* \subset B^*$ is a finite A^*-module (Proposition III.2.3), and consequently we are reduced to show that for every $(f_1, \ldots, f_s) \in A_1^* \times \cdots \times A_s^*$ there is $q \geq 0$ such that

$$\rho^q(f_1, \ldots, f_s) \in A^*.$$

For this, consider the ideal

$$I = \{h \in A^* \mid h(f_1, \ldots, f_s) \in A^*\}.$$

Clearly, we must prove that $\delta \in \sqrt{I}$. But did it not belong, there would be a prime ideal $\mathfrak{q}$ such that $\rho \notin \mathfrak{q}$. Then $\delta' \notin \mathfrak{q}$ and, by *b)* above, $A_{\mathfrak{q}}^*$ would be a local regular ring. In particular it is a domain, and $\mathfrak{q}$ contains one and only one associated prime ideal of (0), say $\mathfrak{q}_1$. Then for all $i > 0$ we find $h_i \in \mathfrak{q}_i \setminus \mathfrak{q}$ and the element $h = \prod_{i>1} h_i \in \bigcap_{i>1} \mathfrak{q}_i \setminus \mathfrak{q}$. We claim that $h \in I$. The reason is that $hf = 0 \bmod \mathfrak{q}_i$ for $f \in A_i^*$ and $i > 1$, which implies

$$h(f_1, \ldots, f_s) = h(f_1, 0, \ldots, 0) = h(f_1, \ldots, f_1) = hf_1 \in A^*.$$

But $I \subset \mathfrak{q}$ and $h \notin \mathfrak{q}$, which is a contradiction. The proof of the lemma is finished. □

We now prove:

Proposition 3.2 *(Zariski's Main Theorem) Let A be a local algebraic ring and A^* a completion of A. Then, A is a normal domain if and only if so is A^*.*

Proof. As usual, to prove the statement for the three completions of A it is enough to do it for the Nash one. Hence we suppose from now on that A^* is the Nash completion of A. One implication is easy: if A^* is integrally closed in its total ring of fractions K^*, so is A in its quotient field K, because $A^* \cap K = A$ (Proposition 1.3 d)). Thus, we assume that A is a normal domain and will prove that A^* is normal, too, or, in other words, that A^* coincides with its integral closure B^* in K^*.

To start with, we prove the following weaker fact:

Let $\mathfrak{P}$ be a height one prime ideal of A and consider the multiplicative system $S_{\mathfrak{P}} = A \setminus \mathfrak{P}$. Then, the ring of fractions $S_{\mathfrak{P}}^{-1} A^*$ is integrally closed in its total ring of fractions, which is K^*.

Indeed, set $D = S_{\mathfrak{P}}^{-1} A^*$. Since A is a domain, no element of $S_{\mathfrak{P}}$ is a zero divisor in A, nor, by Proposition 1.3 c), in A. Thus $A^* \subset D \subset K^*$. Let $z \in K^*$ be integral over D, say $z = a/b$ with $a, b \in A^*$ and b not a zero-divisor in A^*. We consider the ideal

$$I = \{c \in D \,|\, ca \in bD\}.$$

If $1 \in I$ we are done. Otherwise, $I \subset \mathfrak{R}$ for some proper prime ideal $\mathfrak{R}$ of D. Such a prime ideal is in fact a prime ideal of A^* which does not meet S, that is, such that $\mathfrak{R} \cap A \subset \mathfrak{P}$. Moreover, $\mathfrak{R} \cap A \neq (0)$. For, did $\mathfrak{R}$ lie over $(0) \subset A$, it would be an associated prime of $(0) \subset A^*$ (Chevalley's Theorem for Nash completions, Proposition 2.4) and all its elements would be zero divisors in A^*, which $b \in I$ is not. Thus, as $\operatorname{ht}(\mathfrak{P}) = 1$, we conclude $\mathfrak{R} \cap A = \mathfrak{P}$; hence, by Chevalley's Theorem again, $\mathfrak{R}$ is a minimal associated prime of the extended ideal $\mathfrak{P} A^*$. Now, A being a normal domain, the localization $A_{\mathfrak{P}}$ is a normal domain of dimension 1, and consequently its maximal ideal $\mathfrak{P} A_{\mathfrak{P}}$ is principal ([A-McD 9.2]). But

$$\mathfrak{R} D_{\mathfrak{R}} = \mathfrak{R} A^*_{\mathfrak{R}} = \mathfrak{P} A^*_{\mathfrak{R}},$$

since $\mathfrak{R}$ is a minimal prime of $\mathfrak{P} A^*$, and we conclude that the maximal ideal of the local ring $D_{\mathfrak{R}}$ is principal. Since D has no nilpotent element, it follows easily that $D_{\mathfrak{R}}$ is a domain, and by [A-McD 9.2] it is a normal domain. We finally pick any equation of integral dependence of z over D and get after localization at $\mathfrak{R}$ one over $D_{\mathfrak{R}}$. By the preceding remarks, we deduce $z \in D_{\mathfrak{R}}$, which means that there are elements $c, u, v \in D$, $u, v \notin \mathfrak{R}$ with

$$u(va - bc) = 0$$

(as D is not a domain, we have to use carefully the definition of a ring of fractions, see [A-McD §3]). Whence $uv \in I \subset \mathfrak{R}$ and $uv \notin \mathfrak{R}$, contradiction. In this way we have proved our claim.

Now we will show that A^* itself is integrally closed in K^*. By Zariski's Condition D (Lemma 4.1) there is an element $\delta \in A$ such that $\delta B^* \in A^*$. We have the following formula

$$\delta A = \bigcap_{\delta \in \mathfrak{P},\, \operatorname{ht}(\mathfrak{P}) = 1} I(\mathfrak{P}), \tag{1}$$

where $I(\mathfrak{P}) = A \cap \delta A_{\mathfrak{P}}$ for every prime ideal $\mathfrak{P} \subset A$.

Let us admit (1) for the moment, and deduce from it that $B^* = A^*$. Since δ is not a zero divisor in A^*, it suffices to see that $\delta B^* \in \delta A^*$. But, by Proposition 1.3 b), the formula (1) above extends to A^* in the form

$$\delta A^* = \bigcap_{\delta \in \mathfrak{P},\, \operatorname{ht}(\mathfrak{P}) = 1} I(\mathfrak{P})^*,$$

3 Zariski's Main Theorem

and we are reduced to show that

$$\delta B^* \subset I(\mathfrak{P})^*$$

for every height 1 prime $\mathfrak{P} \subset A$ that contains δ. But B^* is integral over $A^* \subset S_\mathfrak{P}^{-1} A^*$, and the latter ring is integrally closed, so that $B^* \subset S_\mathfrak{P}^{-1} A^*$. It follows

$$\delta B^* \subset A^* \cap \delta(S_\mathfrak{P}^{-1} A^*),$$

and we have to see that $A^* \cap \delta(S_\mathfrak{P}^{-1} A^*) \subset I(\mathfrak{P})^*$. Hence let $z \in A^*$ be such that

$$uz - \delta y = 0,$$

where $y \in A^*, u \in S_\mathfrak{P} = A \setminus \mathfrak{P}$ (note that no element of $S_\mathfrak{P}$ is a zero divisor in A^*, which simplifies the description of $S_\mathfrak{P}^{-1} A^*$). We then have the linear equation

$$u\mathbf{z} - \delta \mathbf{y} = 0, \quad c, \delta \in A,$$

with a solution $\mathbf{z} = z$, $\mathbf{y} = y$ in A^*. By Lemma 1.2, for every $\alpha \geq 1$ there is a solution $\mathbf{z} = z_\alpha, \mathbf{y} = y_\alpha$ in A such that $z_\alpha = z \mod \mathfrak{m}^{*\alpha}$. Hence $uz_\alpha = \delta y_\alpha$, and since $c \notin \mathfrak{P}$,

$$z_\alpha \in A \cap \delta A_\mathfrak{P} = I(\mathfrak{P}).$$

Consequently

$$z \in \bigcap_\alpha (I(\mathfrak{P})^* + \mathfrak{m}^{*\alpha}) = I(\mathfrak{P})^*,$$

as wanted.

It only remains to prove the formula (1) stated above. This is a general fact valid in any noetherian normal domain, but as it is not explicitely formulated in our basic reference [A-McD] we include a proof.

Let $\delta A = \mathfrak{a}_1 \cap \cdots \cap \mathfrak{a}_s$ be a primary decomposition of δA, with associated primes

$$\mathfrak{P}_1 = \sqrt{\mathfrak{a}_1}, \ldots, \mathfrak{P}_s = \sqrt{\mathfrak{a}_s}.$$

If we show that

$$\mathrm{ht}(\mathfrak{P}_1) = \cdots = \mathrm{ht}(\mathfrak{P}_s) = 1,$$

then $\mathfrak{a}_j \not\subset \mathfrak{P}_i$ for $j \neq i$, and our formula (1) follows from [A-McD 4.9].

Thus, we put $\mathfrak{P} = \mathfrak{P}_i$, and will see that $\mathfrak{P}$ has height 1, or, more explicitely, that $\mathfrak{P} A_\mathfrak{P}$ is principal. Set $C = A_\mathfrak{P}$ and $\mathfrak{n} = \mathfrak{P} A_\mathfrak{P}$. Then C is a local normal domain with quotient field K. Furthermore, $\mathfrak{n}$ is still an associated prime of δC (again by [A-McD 4.9]). We now consider

$$\mathfrak{n}^{-1} = \{x \in K \,|\, x\mathfrak{n} \subset C\} \supset C,$$

and claim that

$$\mathfrak{n}^{-1}\mathfrak{n} = C \tag{2}$$

Clearly, $\mathfrak{n}^{-1}\mathfrak{n} \subset C$ is an ideal and contains $\mathfrak{n}$. Hence, if it were a proper ideal, it would coincide with $\mathfrak{n}$. It would then follow $\mathfrak{n}(\mathfrak{n}^{-1})^m = \mathfrak{n}$ for all $m \geq 1$, and so

$$a(\mathfrak{n}^{-1})^m \subset C$$

for any chosen non-zero element $a \in \mathfrak{n}$. Hence,

$$C[\mathfrak{n}^{-1}] \subset a^{-1}C.$$

As $a^{-1}C$ is a finite C-module, so would be $C[\mathfrak{n}^{-1}]$, which consequently would be integral over C. As C is normal, we conclude $\mathfrak{n}^{-1} \subset C$. But this is impossible. For, by a general property of associated primes ([A-McD 4.5]), there is $a \in C$ such that $\mathfrak{n}$ is the radical of the ideal $\mathfrak{a} = \{c \in C \mid \delta \text{ divides } ac\}$. Hence $\mathfrak{n}^m \subset \mathfrak{a}$ for some $m \geq 1$. Let $z = a/\delta \in K$; it holds $z\mathfrak{n}^m \subset C$. If $\mathfrak{n}^{-1} \subset C$, then $z \in C$, and we would get $1 \in \mathfrak{a} \subset \mathfrak{n}$, which is absurd. We are done.

Once (2) is proved we are ready to show that $\mathfrak{n}$ is principal. By Nakayama's Lemma, $\mathfrak{n} \neq \mathfrak{n}^2$ and we pick an element $t \in \mathfrak{n} \setminus \mathfrak{n}^2$. Clearly, $t\mathfrak{n}^{-1} \subset C$, and if it were $t\mathfrak{n}^{-1} \subset \mathfrak{n}$ we would get $tC = t\mathfrak{n}^{-1}\mathfrak{n} \subset \mathfrak{n}^2$ by (2). Hence, $t\mathfrak{n}^{-1} = C$, and so $\mathfrak{n} = t\mathfrak{n}^{-1}\mathfrak{n} = tC$ by (2) again.

Thus we have completed the proof that $\mathfrak{n} = \mathfrak{P}A_\mathfrak{P}$ is principal, and with it the proof of (1) and the proof of Zariski's Main Theorem. $\square$

4 Normalization and Completion

We will describe here the normalization of a local algebraic domain. First we have the following consequence of Zariski's Condition D:

Proposition and Definition 4.1 *Let A be a local algebraic domain, K its quotient field and B the integral closure of A in K. Then:*

a) B is a finite A-module and a noetherian ring.

b) B has finitely many maximal ideals, which are the prime ideals $\mathfrak{n}_1, \ldots, \mathfrak{n}_r$ of B lying over the maximal ideal $\mathfrak{m}$ of A.

c) $\sqrt{\mathfrak{m}B} = \mathfrak{n}_1 \cap \cdots \cap \mathfrak{n}_r$.

The ring B is the normalization *of A.*

4 Normalization and Completion

Proof. By Zariski's Condition D (Lemma 3.1), there is a non-zero element $\delta \in A$ such that $\delta B^* \subset A^*$. Hence $\delta B \subset A^* \cap K = A$ (Proposition 1.3 d)). Since A is noetherian, we deduce that δB is a finite A-module, and consequently so is B. It also follows that B is noetherian ([A-McD 6.5]). Let now $\mathfrak{n}$ be a prime ideal of B. By a basic fact on integral dependence ([A-McD 5.8]) $\mathfrak{n}$ is maximal if and only if $\mathfrak{n} \cap A$ is maximal. It follows that the maximal ideals of B are the prime ideals of B lying over $\mathfrak{m}$, or, equivalently, the prime ideals of B containing $\mathfrak{m}B$. Hence they are exactly the associated primes of $\mathfrak{m}B$ and their intersection is the radical of $\mathfrak{m}B$. □

We are ready to compare the normalization of a local algebraic domain and its completion. We recall that our coefficient field is $\mathbb{K} = \mathbb{R}$ or $\mathbb{C}$.

(4.2) Normalization after completion. Let A be a local algebraic domain over $\mathbb{K}$ and K its quotient field. We consider the normalization $B \subset K$ of A and a completion A^* of A.

By Chevalley's Theorem (Proposition 2.1), A^* is reduced and the ideal $(0) \subset A^*$ has finitely many associated primes $\mathfrak{q}_i$. By Proposition and Definition 4.1, B has finitely many maximal ideals $\mathfrak{n}_j$. We are to compare the $\mathfrak{q}_i$'s and the $\mathfrak{n}_j$'s through the normalization B^* of A^*, as described in Proposition III.2.2: B^* is a direct product of the normalizations B_i^* of the domains $A^*/\mathfrak{q}_i$. Recall also that the factors B_i^* are the localizations of B^* at its maximal ideals $\mathfrak{n}_i^*$. As usually, we suppose that A^* is the Nash completion, since what we prove for it will also hold for the others by Nagata's comparison theorems. First of all, we have:

a) $B^* \cap K = B$

Clearly, $B^* \cap K \supset B$. On the other hand, by Zariski's Condition D, there is $\delta \in A$ with $\delta B^* \subset A^*$. Consequently

$$\delta(B^* \cap K) \subset (\delta B^*) \cap K \subset A^* \cap K = A$$

(the last equality by Proposition 1.3 d)). Hence $\delta(B^* \cap K)$, and thus $B^* \cap K$, is a finite A-module. It follows that $B^* \cap K$ is integral over A, and so $B^* \cap K \subset B$. □

Then:

b) Every maximal ideal $\mathfrak{n}_i^*$ of B^* lies over one $\mathfrak{n}_j$ of B.

Since $\mathfrak{n}_i^*$ lies over the maximal ideal of A^*, it lies over the one $\mathfrak{m}$ of A. This implies that $\mathfrak{n}_i^* \cap B$ is a prime ideal of B that lies over $\mathfrak{m}$, and by Proposition and Definition 4.1, it is a maximal ideal of B, that is, some of the $\mathfrak{n}_j$'s. □

Conversely, we have

c) Over every maximal ideal $\mathfrak{n}_j$ of B there lies some maximal ideal $\mathfrak{n}_i^*$ of B^*.

It is enough to show that the extended ideal $\mathfrak{n}_j B^*$ is proper, since then it will be contained in some maximal ideal $\mathfrak{n}_i^*$ of B^*, which of course will lie over $\mathfrak{n}_j$. Now, to see that $\mathfrak{n}_j B^*$ is indeed proper, we argue as follows. As B is a finite A-module, there are elements $z_1, \ldots, z_m \in B$ such that $B = A[z_1, \ldots, z_m]$, and we consider the extensions

$$B \subset C = A^*[z_1, \ldots, z_m] \subset B^*.$$

Since B^* is integral over C, any proper ideal of C extends to a proper ideal of B^* ([A-McD 5.10]) and so we only have to see that $\mathfrak{n}_j C$ is proper. But suppose $1 \in \mathfrak{n}_j C$. Then, there is an expression

$$1 = P_1(a, z)c_1 + \cdots + P_s(a, z)c_s,$$

where $a = (a_1, \ldots, a_p) \in (A^*)^p$, $z = (z_1, \ldots, z_m)$, $c_1, \ldots, c_s \in \mathfrak{n}_j$ and every $P_\ell \in \mathbb{Z}[a, z]$ is linear in the a_k's. We now apply Lemma 1.2 to this equation in the a_k's and find $a' \in A^p$ such that

$$1 = P_1(a', z)c_1 + \cdots + P_s(a', z)c_s.$$

But then $1 \in \mathfrak{n}_j$, which is absurd. This concludes the proof of c). $\square$

(4.3) Completion after normalization. We will next construct the completion of every localization $B_{\mathfrak{n}_j}$ (we keep the notations of 4.2). There are several possibilities, depending on the type of extension we get via the canonical homomorphism $\mathbb{K} = A/\mathfrak{m} \subset B/\mathfrak{n}_j = \mathbb{K}'$. Note that since B is integral over A, that extension is algebraic, and must be either $\mathbb{R} \subset \mathbb{R}$, $\mathbb{R} \subset \mathbb{C}$ or $\mathbb{C} \subset \mathbb{C}$. The simplest situation is:

a) The extension is trivial, that is, $\mathbb{K} = B/\mathfrak{n}_j$.

We pick a surjective homomorphism $\mathbb{K}_0[\mathbf{x}] \to A$, with $\mathbf{x} = (\mathbf{x}_1, \ldots, \mathbf{x}_n)$, and elements $z_1, \ldots, z_m$ in B such that $B = A[z_1, \ldots, z_m]$. Since the extension $\mathbb{K} \subset B/\mathfrak{n}_j$ is trivial, there are $z_1^0, \ldots, z_m^0 \in \mathbb{K} \subset B$ with

$$z_1 - z_1^0, \ldots, z_m - z_m^0 \in \mathfrak{n}_j.$$

We then put $\mathbf{y} = (\mathbf{y}_1, \ldots, \mathbf{y}_m)$ and extend $\mathbb{K}_0[\mathbf{x}] \to A$ to a surjection

$$\mathbb{K}_0[\mathbf{x}, \mathbf{y}] \to B_{\mathfrak{n}_j}$$

by

$$\mathbf{y}_1 \mapsto z_1 - z_1^0, \ldots, \mathbf{y}_m \mapsto z_m - z_m^0.$$

We have thus shown that $B_{\mathfrak{n}_j}$ is a local algebraic ring over $\mathbb{K}$, and can consider its Nash completion $(B_{\mathfrak{n}_j})^*$.

Now, since the canonical homomorphism $A \to B_{\mathfrak{n}_j}$ is local, it extends to a Nash homomorphism $A^* \to (B_{\mathfrak{n}_j})^*$ (this follows immediately from Proposition II.1.3). The kernel of this extension lies over $(0) \subset A$, because $A \to B_{\mathfrak{n}_j}$ is injective. Then, by Chevalley's Theorem for Nash completions (Proposition 2.4), that kernel is an associated prime $\mathfrak{q}_k$.

Next, we consider a maximal ideal $\mathfrak{n}_i^*$ of B^* that lies over $\mathfrak{n}_j$ and the corresponding local homomorphism $B_{\mathfrak{n}_j} \to B_i^* = B_{\mathfrak{n}_i^*}^*$. Again we can extend this to a Nash homomorphism $(B_{\mathfrak{n}_j})^* \to B_i^*$ and we obtain

$$A^* \to A^*/\mathfrak{q}_k \hookrightarrow (B_{\mathfrak{n}_j})^* \to B_i^*.$$

4 Normalization and Completion   123

Clearly, this is the canonical homomorphism $A \to B_i^*$, and $\mathfrak{q}_k$ is its kernel; in particular, $k = i$. Moreover, by Zariski's Main Theorem (Proposition 3.2), the Nash ring $(B_{\mathfrak{n}_j})^*$ is normal, and we conclude that $(B_{\mathfrak{n}_j})^* = B_i^*$. This shows in addition that the maximal ideals $\mathfrak{n}_i^*$ lying over $\mathfrak{n}_j$ correspond to the associated primes $\mathfrak{q}_i$ which occur as kernels of Nash homomorphisms $A \to (B_{\mathfrak{n}_j})^*$ that extend the canonical inclusion $A \to B_{\mathfrak{n}_j}$. But such a Nash homomorphism is completely determined by the images of the generators of the maximal ideal $\mathfrak{m}^*$ of A^* (Proposition I.1.3 a)), and since $\mathfrak{m}^* = \mathfrak{m}A^*$, those images are prescribed from the very beginning. Whence, there is a unique $\mathfrak{n}_i^*$ lying over $\mathfrak{n}_j$. □

The other possibility is that $\mathbb{K} = \mathbb{R} \subset B/\mathfrak{n}_i = \mathbb{C}$. Then if $\mathfrak{n}_i^*$ lies over $\mathfrak{n}_j$, the coefficient field of $B_i^* = B_{\mathfrak{n}_i^*}^*$ must be $\mathbb{C}$. We still need to distinguish two subcases:

b) B is not 2-real.

Then $\sqrt{-1} \in K$ (Proposition and Definition II.5.6), and since B is integrally closed in K, $\sqrt{-1} \in B$ and $\mathbb{C} \subset B$. We now repeat the reasoning of *a)*, but notice that $z_1^0, \ldots, z_m^0 \in \mathbb{C}$. This leads to a surjection $\mathbb{C}_0[\mathbf{x}, \mathbf{y}] \to B_{\mathfrak{n}_j}$. In this way, $B_{\mathfrak{n}_j}$ is a local algebraic ring over $\mathbb{C}$. The rest of the argument goes the same: we get a Nash homomorphism $A^* \to (B_{\mathfrak{n}_j})^*$ whose kernel is the associated prime $\mathfrak{q}_i$ corresponding to $\mathfrak{n}_i^*$. From this we conclude that there is a unique maximal ideal $\mathfrak{n}_i^*$ lying over $\mathfrak{n}_j$ and the localization B_i^* of B^* at $\mathfrak{n}_i^*$ is the Nash completion $(B_{\mathfrak{n}_j})^*$ of $B_{\mathfrak{n}_j}$. It is important to observe the extension of the coefficient field along this process, and the small abuse of terminology when we consider homomorphisms from A^*, whose coefficient field is $\mathbb{R}$, to $(B_{\mathfrak{n}_j})^*$, whose coefficient field is $\mathbb{C}$. Clearly this abuse does not affect the argument. □

c) B is 2-real.

Then $\sqrt{-1} \notin K$, and the extension $B[\sqrt{-1}]$ is a domain (Proposition and Definition II.5.6). As the coefficient field of the Nash ring B_i^* is $\mathbb{C}$, we get a homomorphism $B[\sqrt{-1}] \to B_i^*$. Hence $\mathfrak{n}_i^*$ lies over a maximal ideal $\mathfrak{n}'$ of $B[\sqrt{-1}]$ which in turn lies over $\mathfrak{n}_j$. We can again construct a surjective homomorphism $\mathbb{C}_0[\mathbf{x}, \mathbf{y}] \to B[\sqrt{-1}]_{\mathfrak{n}'}$ and $B[\sqrt{-1}]_{\mathfrak{n}'}$ is a local algebraic ring over $\mathbb{C}$. The Nash completion of this local algebraic ring is still denoted by $(B_{\mathfrak{n}_j})^*$ and once again we obtain a Nash homomorphism $A^* \to (B_{\mathfrak{n}_j})^*$ whose kernel determines $\mathfrak{n}_i^*$. Also, $(B_{\mathfrak{n}_j})^*$ is canonically isomorphic to B_i^*.

The more delicate point here is the uniqueness of $\mathfrak{n}_i^*$. As in the previous cases, this is a consequence of the fact that the kernel $\mathfrak{q}_i$ of the Nash homomorphism $\phi : A^* \to (B_{\mathfrak{n}_j})^* = B_i^*$ is completely determined by its restriction $\varphi : A \to B_{\mathfrak{n}_j}$ to A. This fact in turn comes from the following characterization of $\mathfrak{q}_k$. Let $f \in A^*$ and for every $\alpha \geq 0$ write f in the form

$$f = f_\alpha + h_\alpha, \ f_\alpha \in A, \ h_\alpha \in \mathfrak{m}^{*\alpha}.$$

Then $\phi(f) = 0$ if and only if for every $\beta \geq 0$ there is $\alpha \geq \beta$ such that $\varphi(f_\alpha) \in \mathfrak{n}_j^\beta$.

To obtain this characterization, suppose first that the condition holds true. Then for every $\beta \geq 0$ there is $\alpha \geq \beta$ with

$$\phi(f) = \varphi(f_\alpha) + \phi(h_\alpha) \in \mathfrak{n}_j^\beta B_i^* + \mathfrak{n}_i^{*\alpha} \subset \mathfrak{n}_i^{*\beta},$$

and so

$$\phi(f) \in \bigcap_\beta \mathfrak{n}_i^{*\beta} = (0).$$

Conversely, suppose that $\phi(f) = 0$. Then

$$\varphi(f_\alpha) = -\phi(h_\alpha) \in \mathfrak{n}_i^{*\alpha},$$

and we must see that for every $\beta \geq 0$ there is some α such that

$$\mathfrak{n}_i^{*\alpha} \cap B_{n_j} \subset \mathfrak{n}_j^\beta.$$

We recall the construction of $(B_{n_j})^* = B_i^*$: it is the Nash completion of the local algebraic ring $B[\sqrt{-1}]_{\mathfrak{n}'}$. By flatness (Proposition 1.3 a))

$$(\mathfrak{n}_i^*)^\alpha \cap B[\sqrt{-1}]_{\mathfrak{n}'} = \mathfrak{n}'^\alpha,$$

and consequently we are reduced to find α such that

$$\mathfrak{n}'^\alpha \cap B_{n_j} \subset \mathfrak{n}_j^\beta.$$

Since all prime ideals of $B[\sqrt{-1}]$ lying over $\mathfrak{n}_j$ are maximal (a basic property of integral dependence, [A-McD 5.8]) we deduce that $\mathfrak{n}'$ is the unique prime ideal of $B[\sqrt{-1}]_{\mathfrak{n}'}$ lying over $\mathfrak{n}_j$. This means that

$$\mathfrak{n}' = \sqrt{\mathfrak{n}_j B[\sqrt{-1}]_{\mathfrak{n}'}},$$

and since this localization is a noetherian ring, we get

$$\mathfrak{n}'^\gamma \subset \mathfrak{n}_j B[\sqrt{-1}]_{\mathfrak{n}'}$$

for some $\gamma \geq 1$. Hence

$$\mathfrak{n}'^{\gamma\beta} \subset \mathfrak{n}_j^\beta B[\sqrt{-1}]_{\mathfrak{n}'},$$

and we will have finished if

$$\mathfrak{n}_j^\beta B[\sqrt{-1}]_{\mathfrak{n}'} \cap B_{n_j} = \mathfrak{n}_j^\beta. \tag{1}$$

To see this, consider generators $h_1, \ldots, h_m$ of $\mathfrak{n}_j^\beta$. An element a belongs to the left hand side of (1) if and only if

$$ua = c_1 h_1 + \cdots + c_m h_m,$$

where $u, c_\ell \in B[\sqrt{-1}]$, $u \notin \mathfrak{n}'$. We then write

$$u = u' + \sqrt{-1}u'', \quad c_\ell = c'_\ell + \sqrt{-1}c''_\ell,$$

where $u', u'', c'_\ell, c''_\ell \in B$. As $u \notin \mathfrak{n}'$, then either $u' \notin \mathfrak{n}'$ or $u'' \notin \mathfrak{n}'$. Suppose, say, the first condition. Then $u' \notin \mathfrak{n}_j$ and we have

$$u'a = c'_1 h_1 + \cdots + c'_m h_m.$$

So a belongs to the ideal generated by $h_1, \ldots, h_m$ in $B_{\mathfrak{n}_j}$, which is the right hand side of (1). This gives the non-immediate inclusion and concludes the proof of the characterization of the elements of $\mathfrak{q}_k$. As explained before, the uniqueness of $\mathfrak{n}_i^*$ is a consequence of that characterization, and the discussion of this case $c)$ is complete. □

We summarize the constructions of the preceding two paragraphs in a single statement:

Proposition 4.4 *Let A be a local algebraic domain, A^* a completion of A and B the normalization of A. Then, the normalization of A^* is the completion of B, that is,*

$$(A^*)^\nu \simeq (B_{\mathfrak{n}_1})^* \times \cdots \times (B_{\mathfrak{n}_s})^*,$$

where $\mathfrak{n}_1, \ldots, \mathfrak{n}_s$ are the maximal ideals of B. In particular, the isomorphisms

$$(A^*/\mathfrak{q}_i)^\nu \simeq (B_{\mathfrak{n}_i})^*, \ 1 \leq i \leq s,$$

give a bijection between the associated primes $\mathfrak{q}_i$ of $(0) \subset A^*$ and the maximal ideals $\mathfrak{n}_i$ of B.

5 Efroymson's Theorem

The goal of this section is to describe the behaviour of the completions of a local algebraic ring with respect to orderings. To do this we fix the following notations.

Let A be a local algebraic domain over $\mathbb{R}$, $\mathfrak{m}$ its maximal ideal and K its quotient field. Let A^* denote the Nash (resp. analytic, formal) completion of A. Namely, we have $A = \mathcal{R}_n/\mathfrak{p}$, $A^* = \mathcal{R}_n^*/\mathfrak{p}^*$. By Chevalley's Theorem (Proposition 2.1) the ideal $\mathfrak{p}^*$ is radical, and its associated primes have all height $= \text{ht}(\mathfrak{p})$, say

$$\mathfrak{p}^* = \mathfrak{q}_1 \cap \cdots \cap \mathfrak{q}_s.$$

Then, for every $i = 1, \ldots, s$, we put $A_i^* = A^*/\mathfrak{q}_i$ and denote by K_i^* the quotient field of A_i^*.

In this situation, the problem is to know how an ordering of the domain A can be extended to some A_i^*. For this we need a new notion:

Definition 5.1 *Let B be a local domain, with maximal ideal $\mathfrak{n}$. Then:*

a) A local ordering *of B is an ordering $<$ such that for every $g \in B$ and $f \in \mathfrak{n}$ with $0 < g < f$ it follows $g \in \mathfrak{n}$; we also say that $\mathfrak{n}$ is* convex *with respect to $<$.*

b) We say that B is locally real *if there exists some local ordering of B.*

Remarks 5.2 Let B be a local algebraic (resp. a Nash, an analytic, a formal) ring over $\mathbb{R}$ with maximal ideal $\mathfrak{n}$, and $<$ a local ordering of B. Then:

a) Any $h \in B$ with $h(0) > 0$ is positive in $<$.

If h were negative, then $-h$ would be positive. Hence

$$0 < -h < h(0) - h, \ h(0) - h \in \mathfrak{n}, \ -h \notin \mathfrak{n},$$

and $\mathfrak{n}$ would not be convex. □

b) Every $h \in B$ is bounded by a real number, that is, $-M < h < M$ for some positive real number M.

Take $-M < h(0) < M$. Then $(M - h)(0) = M - h(0) > 0$ and by *a)*, $M - h > 0$. Analogously, $h + M > 0$, and the assertion is proved. □

c) The maximal ideal $\mathfrak{n}$ consists exactly of the *infinitesimal* elements, that is, the elements $f \in B$ such that $-\varepsilon < f < \varepsilon$ for every positive real number ε.

If $f \in \mathfrak{n}$, then $(\varepsilon - f)(0) = (\varepsilon + f)(0) = \varepsilon > 0$ and by *a)*, $-\varepsilon < f < \varepsilon$. If $f \notin \mathfrak{n}$, then $f(0) \neq 0$ and we set $\varepsilon = \frac{1}{2}f(0) > 0$. In case $f(0) > 0$, $(f - \varepsilon)(0) = \frac{1}{2}f(0) > 0$ and, from *a)* once more, it follows that $0 < \varepsilon < f$. A similar argument shows that $f < -\varepsilon < 0$ in case $f(0) < 0$. Thus f is not infinitesimal. This concludes the proof. □

When the Implicit Functions Theorem holds, all orderings are local:

Proposition 5.3 *Let B be a Nash (resp. an analytic, a formal) ring over $\mathbb{R}$ with maximal ideal $\mathfrak{n}$. Then every ordering $<$ of B is a local ordering.*

Proof. Suppose $0 < g < f$ with $g \in B$ and $f \in \mathfrak{n}$. Then $f(0) = 0$ and we must see that $g(0) = 0$. If $g(0) < 0$, then $-g = h^2$ for some $h \in B$ (II.4.7) and so $-g > 0$, against the hypothesis. Thus $g(0) \geq 0$. If $g(0) > 0$, then $(f - g)(0) = -g(0) < 0$ and $f - g = -h^2$ for some $h \in B$ (II.4.7). Hence $f - g < 0$, also against the hypothesis. Consequently, $g(0) = 0$ as wanted. □

After this preparation we can state the central result:

Proposition 5.4 *Let $<$ be an ordering of A. The following assertions are equivalent:*

  a) $<$ *is a local ordering.*

  b) $<$ *extends to some $A_i^* = A^*/\mathfrak{q}_i$.*

If this is the case, $\mathfrak{q}_i$ corresponds via the bijection of Proposition 4.4 to the unique maximal ideal $\mathfrak{n}_i$ of the normalization B of A which is convex with respect to $<$.

Proof. The implication $b) \Rightarrow a)$ is an immediate consequence of Proposition 5.3. For the converse implication it is enough to prove the case of the Nash completion, by Proposition V.4.9 *c)*, and to do it, we start with the particular case $A = \mathcal{R}_d$, $A^* = \mathcal{N}_d$.

5 Efroymson's Theorem

We will use Serre's Criterion as stated in the proof of Proposition V.4.9 e): let $f_1, \ldots, f_m \in \mathcal{R}_d$ be positive in $<$ and let us see that the equation

$$f_1 \mathbf{y}_1^2 + \cdots + f_m \mathbf{y}_m^2 = 0 \tag{1}$$

has in $\mathcal{N}_d$ only the trivial solution.

Indeed, for every real number $\varepsilon > 0$ the polynomial

$$h_\varepsilon = \varepsilon - (\mathbf{x}_1^2 + \cdots + \mathbf{x}_d^2)$$

is positive in $<$ (because $h_\varepsilon(0) > 0$ and $<$ is local). Then, by *E. Artin's Specialization Theorem* ([L XI.3 Lem.2]), we have

$$\{x \in \mathbb{R}^d \,|\, h_\varepsilon(x) > 0, f_1(x) > 0, \ldots, f_m(x) > 0\} \neq \emptyset.$$

As this holds for every $\varepsilon > 0$ we deduce that the origin is adherent to the open set

$$Z = \{x \in D \,|\, f_1(x) > 0, \ldots, f_m(x) > 0\},$$

where $D \subset \mathbb{R}^d$ is an open polycylinder centered at the origin on which $f_1, \ldots, f_m$ are well defined analytic functions (such a D exists because the f_i's are rational functions whose denominators do not vanish at the origin).

Let now $g_1, \ldots, g_m \in \mathcal{N}_n \subset \mathcal{O}_n$ be a solution of the equation (1), and consider the associated functions

$${}^a g_i : D(g_i) \to \mathbb{R}, \; 1 \leq i \leq m$$

(see I.2). Since $f_1 g_1^2 + \cdots + f_m g_m^2 = 0$ in $\mathcal{O}_d$, this holds the same for the associated functions on some non-empty open neighborhood of the origin

$$U \subset D \cap D(g_1) \cap \cdots \cap D(g_m).$$

It follows that each ${}^a g_i$ vanishes on the open set $U \cap Z \subset D(g_i)$. But Z is adherent to the origin, so that $U \cap Z \neq \emptyset$, and from the Identity Principle (Proposition I.2.9) we deduce $g_i = 0$. This ends the proof of the particular case.

We next apply to $\mathfrak{p}$ Noether's Projection Lemma for polynomials ([A-McD §5 Ex.16]), and can assume the following conditions:

a) *The canonical homomorphism $\mathcal{R}_d \to A$ is local and injective, with $d = \dim(A)$.*

b) *The classes $\theta_j = \mathbf{x}_j \bmod \mathfrak{p} \in A$ are integral elements over $\mathcal{R}_d$.*

Let now $<$ be our local ordering of A. It restricts to an ordering of $\mathcal{R}_d$, which is also local; this restriction is again denoted by $<$. By the particular case already solved, $<$ extends from $\mathcal{R}_d$ to $\mathcal{N}_d$, or, in other words, from the quotient field L of $\mathcal{R}_n$ to the one L^* of $\mathcal{N}_d$. Then, we consider the real closure F of L^* and get a chain of ordered algebraic extensions

$$L \subset L^* \subset F.$$

In addition, we have the ordered finite extension $L \subset K$, that must also embed in F: $K \subset F$. For these facts we refer to [L XI.2 Th.3].

On the other hand, consider the element θ_j and a monic polynomial $P \in \mathcal{R}_d[t]$ such that $P(\theta_j) = 0$ (which exists by b)). By Weierstrass's Preparation Theorem

$$P = UP_j, \quad U, P_j \in \mathcal{N}_d[t], \quad U(0,0) = \eta \neq 0,$$

and P_j is distinguished. Set $\theta_j^* = \jmath(\theta_j)$. Since $<$ is local, the element $\theta_j \in \mathfrak{m}$ is infinitesimal (Remark 5.2 c)), that is,

$$-\varepsilon < \theta_j < \varepsilon \quad \text{for every real number } \varepsilon > 0.$$

Suppose now $U(\theta_j) = 0$. Then

$$\theta_j^q + a_1 \theta_j^{q-1} + \cdots + a_q = 0, \quad a_q(0) = U(0,0) = \eta \neq 0,$$

where the coefficients $a_\ell \in \mathcal{N}_d$ are bounded (Remark 5.2 b)):

$$-M < a_\ell < M \quad \text{for some real number } M > 1.$$

A straightforward computation gives

$$\mp a_q = \pm(\theta_j^{q-1} + a_1 \theta_j^{q-2} + \cdots + a_{q-1})\theta_j < qM\varepsilon \tag{2}$$

for every $\varepsilon > 0$.

Then if, say, $\eta > 0$, we have

$$(a_q - qM\varepsilon)(0) = \eta - qM\varepsilon > 0$$

for small enough $\varepsilon > 0$, and by Remark 5.2 a)

$$a_q > qM\varepsilon,$$

against (2). If $\eta < 0$ we would similarly conclude $-a_q > qM\varepsilon$ also against (2).

The conclusion is that θ_j is a root of the distinguished polynomial P_j.

Once we have the P_j's, we use the method of the second half of the proof of Lemma III.1.1 to construct a homomorphism

$$\varphi: \mathcal{N}_n \to \mathcal{N}_d[\theta_{d+1}, \ldots, \theta_n] \subset F$$

such that

$$\mathbf{x}_1 \mapsto \mathbf{x}_1, \ldots, \mathbf{x}_d \mapsto \mathbf{x}_d; \mathbf{x}_{d+1} \mapsto \theta_{d+1}, \ldots, \mathbf{x}_n \mapsto \theta_n.$$

By construction, $\varphi|\mathcal{R}_n$ coincides with the composition

$$\mathcal{R}_n \to \mathcal{R}_n/\mathfrak{p} = A \subset K \subset F,$$

and so $\mathfrak{p} = \mathcal{R}_n \cap \ker(\varphi)$. It follows from Proposition 2.4 that $\ker(\varphi)$ is an associated prime $\mathfrak{q}_i$ of $\mathfrak{p}^*$, and we get another embedding

$$A_i^* = \mathcal{N}_n/\mathfrak{q}_i \hookrightarrow F.$$

Finally, the restriction to A_i^* of the ordering of the real closed field F gives the extension of $<$ to A_i^* we were looking for. This completes the equivalence between *a)* and *b)*.

We now prove the last assertion of the statement. The bijection of Proposition 4.4 gives a maximal ideal $\mathfrak{n}_i$ of B and a local inclusion $B_{\mathfrak{n}_i} \to B_i^*$, where B_i^* stands for the normalization of A_i^*. The extension of $<$ to A_i^* is actually an ordering of the quotient field of A_i^*, and, consequently, an ordering of B_i^*. Thus the coefficient field of B_i^* is $\mathbb{R}$ (Proposition III.2.6), and by Proposition 5.3 this ordering of B_i^* is local. Since the inclusion $B_{\mathfrak{n}_i} \to B_i^*$ is local, the restriction to $B_{\mathfrak{n}_i}$ is also a local ordering. But this latter restriction is exactly our initial ordering $<$ of A, because B is contained in the quotient field K of A. Thus we see that $\mathfrak{n}_i$ is convex with respect to $<$. It only remains to see that no other $\mathfrak{n}_j$ is convex with respect to $<$. But, if $j \neq i$, there are elements

$$x \in \mathfrak{n}_j \setminus \mathfrak{n}_i,\ y \in \mathfrak{n}_i \setminus \mathfrak{n}_j,$$

and the element $z = x^2 - y^2 \in B$ verifies

$$x^2 - z = y^2 \in \mathfrak{n}_i \setminus \mathfrak{n}_j,\ y^2 + z = x^2 \in \mathfrak{n}_j \setminus \mathfrak{n}_i.$$

Since $\mathfrak{n}_i$ is convex with respect to $<$, the first condition implies $z > 0$. Indeed, if $z < 0$ then $0 < x^2 < x^2 - z$ and by convexity $x^2 \in \mathfrak{n}_i$, contradiction. Similarly, if $\mathfrak{n}_j$ were also convex, the second condition would imply $z < 0$, which is impossible. We are thus done. $\square$

From the preceding result and the fact that all orderings of the A_i^*'s are local (Proposition 5.3) it follows immediately:

Corollary 5.5 *(Efroymson) A local algebraic domain is locally real if and only if some A_i^* is formally real.*

Bibliographical Note

1. The contents of I.1 and I.2 are standard. We have been inspired by [Abhyankar] and [Grauert Fritzsche]; also by [Cartan 1962].
2. The proof of the Division Theorem given in III.3 is adapted from [Gersten], with the modifications required to work in the convergent case. A similar (but not so short) proof appears in [Tougeron].
3. Mather's Finiteness Theorem (II.1) and Noether's Projection Lemma (II.2) follow [Abhyankar] and [Tougeron], with the changes needed to avoid quotations from Commutative Algebra.
4. The construction presented in II.3 of a regular system of parameters of a prime ideal of power series is close to the one given in [Abhyankar], with some differences motivated by later technical needs. Of course, we have also used [Gunning Rossi].
5. The treatment of Jacobians given in II.4 follows that of [Tougeron]
6. The discussion of complexification presented in II.5 should be considered folklore, although we do not know any complete reference. The best one would be [Cartan 1957].
7. The study of integral closures (III.1) and normalizations (III.2) comes essentially from [Abhyankar] and [Tougeron]. The details concerning the real case and complexifications are again folklore, without previous explicit reference that we know.
8. The treatment of uniformization in dimension 1 given in III.3 is surely well-known to specialists. We have tried a most straightforward approach, with some hidden mention to valuation theory.
9. Our proof of Newton-Puiseux's Theorem is also well-known to specialists. The standard reference is [Walker]; the real case is explicitely formulated in [Lassalle].
10. The use of parametrizations as points and the subsequent proof of Rückert's Nullstellensatz (IV.1 and IV.2) appears in [Tougeron]. It is connected to the model theoretic approach of [Robinson].
11. The Homomorphism Theorem and the Real Nullstellensatz obtained in IV.3 and IV.4 are inspired in [Lassalle]. In fact, this is a reformulation of the original ideas of [Risler], with some simplifications. In addition, the results are a little stronger than in the prior literature.
12. The same can be said of the solution to Hilbert's 17th Problem described in IV.5. Here we have also robbed some ideas from [Stengle].
13. The qualitative and quantitative remarks in IV.5 concerning sums of squares come from [Bochnak Risler] and [Motzkin]. In the first case our proof is different, and in the second we also apply some trick from [Choi Lam].
14. Sections V.1 and V.2 are based on [Tougeron]. We have discarded the technicalities as much as possible. The discussion of the equivalence problem (V.2) is done along the same lines as [Shiota].
15. Algebraic power series and Nash rings are introduced and studied in V.5 following [Lazzeri Tognoli]. They provide an intermediate step between algebraic

and analytic geometry, which is often useful to connect both fields.

16. The proofs of Chevalley's Theorem, Zariski's Main Theorem and the comparison of completions and normalizations of VI.2, VI.3 and VI.4 profit from this idea: by means of algebraic power series and Jacobian Criteria we can simplify Zariski's classical arguments ([Zariski Samuel]).

17. The same remark applies to our proof of Efroymson's Theorem in VI.5, which is simpler than the original one in [Efroymson]. Furthermore, this approach leads to a better result and describes more accurately the situation in the real case.

Finally, we would like to comment on the names chosen for many results in the book. They fall in either of the following cases: authors of a statement close to ours, authors of the ideas that led naturally to our formulation, or authors of a framework in which our presentation fits comfortably.

The quotations above refer to the following books and papers:

S. S. Abhyankar: Local Analytic Geometry, *Pure and Applied Mathematics* **XIV**, Academic Press: New York-London 1964.

J. Bochnak, J.-J. Risler: Le théorème des zéros pour les variétés analytiques réelles de dimension 2, *Ann. Sc. Éc. Norm. Sup.* 4e serie, **8** (1975) 353-364.

H. Cartan: Variétés analytiques réelles et variétés analytiques complexes, *Bull. Soc. Math. France* **85** (1957) 77-99.

H. Cartan: Théorie elementaire des fonctions analytiques d'une ou plusieurs variables complexes, *Enseignement des Sciences* **1**, Hermann: Paris 1962.

M.D. Choi, T.Y. Lam: Extremal positive definite forms, *Math. Ann.* **231** (1977) 1-18.

G.W. Efroymson: Local reality on algebraic varieties, *J. Algebra* **29** (1974) 133-142.

S.M. Gersten: A short proof of the algebraic Weierstrass Preparation Theorem, *Proc. A.M.S.* **88** (1983) 751-752.

H. Grauert, K. Fritzsche: Several complex variables, *Graduate Texts in Mathematics* **38**, Springer-Verlag: New York-Heidelberg-Berlin 1976.

R.C. Gunning, H. Rossi: Analytic functions of several complex variables, *Series in Modern Analysis*, Prentice-Hall: Englewood Cliffs 1965.

G. Lassalle: Sur le théorème the zéros différentiables, *in* Singularités d'applications différentiables, *Lecture Notes in Mathematics* **535**, Springer-Verlag: Berlin-Heidelberg-New York 1975.

F. Lazzeri, A. Tognoli: Alcune proprieta degli spazi algebrici, *Ann. Scuola Norm. Sup. Pisa* **24** (1970) 597-632.

T.S. Motzkin: The arithmetic-geometric inequality, *in* Inequalities, Academic Press: New York-London 1967.

J.-J. Risler: Les théorèmes des zéros en géometries algébrique et analytique réelles, *Bull. Soc. Math. France* **104** (1976) 113-127.

A. Robinson: Germs, *in* Applications of Model Theory, Holt-Rinehart-Winston: New York 1969.

M. Shiota: On the equivalence of differentiable mappings and analytic mappings, *Publ. Math. I.H.E.S.* **54** (1981) 37-122.

G. Stengle: A Nullstellensatz and a Positivstellensatz in semialgebraic geometry, *Math. Ann.* **207** (1974) 87-97.

J.-C. Tougeron: Idéaux des fonctions différentiables, *Ergebnisse der Mathematik (2)* **71**, Springer-Verlag: Berlin-Heidelberg-New York 1972.

R.J. Walker: Algebraic curves, Springer-Verlag: Berlin-Heidelberg-New York 1978.

O. Zariski, P. Samuel: Commutative Algebra I, II, *Graduate Texts in Mathematics* **28**, **29**, Springer-Verlag: New York-Heidelberg-Berlin 1980.

Index

Abhyankar's Parametrization 29
Algebraic power series 106
Algebraicity of an isolated hypersurface singularity 92
Analytic completion of a local algebraic ring 111
Analytic completion of a Nash ring 109
Analytic homomorphism 16
Analytic ring 16
Analyticity of integral closures 49
Analyticity of the function associated to a convergent power series 10
Annihilator of a module 38
Artin-Schreier's Criterion 80

Birational equivalence to a hypersurface 30
Branches of a 1-dimensional reduced analytic ring 53

Chain Rule 9
Chevalley's Theorem 112
Classical Implicit Functions Theorem 36
Classical Inverse Function Theorem 36
Coefficient field 16, 109, 110
Completion of a local ordering 125ff
Completion of the normalization of a local algebraic domain 122ff
Completion of the normalization of an analytic ring 104
Complexification 39
Conjugation 39ff
Contractive map 12
Convergent power series 4
Convergent Puiseux series 59
Convex ideal 125

Derivatives of an algebraic power series 107
Derivatives of a power series 9ff
Distinguished polynomial 11
Domain of a power series 4

Efroymson's Theorem 129
Equidimensionality Jacobian Criterion 37
Equivalent power series 91

Factorization of a distinguished polynomial 22
Finite homomorphism 18

Finiteness of integral closures 27
Fixed Point Theorem 13
Flatness 100, 112
Formal completion of a local algebraic ring 111
Formal completion of a Nash ring 109
Formal completion of an analytic ring 99
Formal homomorphism 16
Formal power series 4
Formal Puiseux series 58
Formal ring 16
Formally real domain 73
Function associated to a convergent power series 5
Function associated to a formal power series 64

Generic changes of coordinates 25
Geometric characterization of dimension in the complex case 70
Geometric characterization of finiteness in the complex case 70

Height of an ideal 22
Hensel's Lemma 14
Hessian matrix 72
Hilbert's 17th Problem for power series with real coefficients 81
Homomorphism Theorem 74
Hypersurface singularity 71

Identity Principle 11
Infinitesimal element 64, 126
Invertible formal power series 7
Invertible convergent power series 9
Isolated hypersurface singularity 71
Isolated singularity 67
Iterated series 3

Jacobian ideal 31

Leibnitz Formula 9
Local algebraic homomorphism 110
Local algebraic ring 110
Local ordering 125
Local Parametrization Theorem 35
Locally real domain 126

M. Artin's Approximation Theorem 94
Mather's Finiteness Theorem 18
Milnor number 72
Morse's Lemma 93
Morse singularity 72
Motzkin's counterexample 84
Multiplicity computed through normalization 54ff
Multiplicity of a 1-dimensional reduced analytic ring 53
Multiplicity of a planar ring 56

Nagata's comparison results 101ff
Nash completion of a local algebraic ring 111
Nash homomorphism 109
Nash ring 109
Newton-Puiseux's Theorem 61
Noether's Projection Lemma 24
Normal analytic (resp. formal) ring 49
Normalization of the complexification of a reduced analytic (resp. formal) ring over the reals 52
Normalization of a local algebraic domain 120
Normalization of a reduced analytic ring 49, 51
Normalization of a 1-dimensional reduced analytic ring 54

Operations with power series 6ff
Order of a power series 4
Ordering of a domain 73
Ordering of the ring of Puiseux series over the reals 73

Positive semidefinite power series 81
Positive semidefinite power series in two indeterminates 85
Primitive Element Theorem 24

Quadratic transform 57ff
Quasifinite homomorphism 18

2-real domain 42
Real radical 79
Reduced ring 41
Regular point of dimension 1 of a zero set 67
Regular power series 11
Regular systems of parameters of a ring of power series 29, 35
Regularity ideal 31
Regularity Jacobian Criterion 32
Risler's Nullstellensatz 78
Roots of invertible power series 37
Rückert's Division Theorem 11
Rückert's Division Theorem for algebraic power series 108
Rückert's Nullstellensatz 68
Rückert's Parametrization 27ff

Schwartz Rule 9
Serre's Criterion 105
Singular point of dimension 1 of a zero set 67
Standard Sturm's sequence 77
Sturm's Theorem 77
Substitution of power series 7
Substitution of algebraic power series 106ff
Sum of a series 1
Summable family of power series 6

Taylor Expansion 9
Tougeron's Implicit Functions Theorem 89
Transversal changes of coordinates 22ff

Uniform convergence of a power series 4

Weierstrass's Preparation Theorem 14
Whitney's Umbrella 78

Zariski's Condition D 115
Zariski's Main Theorem 117
Zero set 64
Zero ideal 65